North on the Wing

North on the Wing

TRAVELS WITH THE SONGBIRD MIGRATION OF SPRING

BRUCE M. BEEHLER

ILLUSTRATIONS BY JOHN T. ANDERTON

 Smithsonian Books   WASHINGTON, DC

This book may be purchased for educational, business, or sales promotional use.
For information, please write Special Markets Department, Smithsonian Books,
P.O. Box 37012, MRC 513, Washington, DC 20013

Published by Smithsonian Books
Director: Carolyn Gleason
Creative Director: Jody Billert
Managing Editor: Christina Wiginton
Project Editor: Laura Harger
Editorial Assistant: Jaime Schwender
Designed by Carol Beehler
Typeset by Scribe Inc.
Endpaper maps by Bill Nelson

ISBN 978-1-58834-613-1 cloth
ISBN 978-1-58834-614-8 e-book

Manufactured in the United States of America
22 21 20 19 18   5 4 3 2 1

TITLE PAGE: Scissor-tailed Flycatchers

In memory of Cary Baxter Beehler,

Edwin Way Teale, and Nellie Donovan Teale

CONTENTS

WOOD WARBLERS OF EASTERN NORTH AMERICA
A Life List

The warblers are listed below in systematic order, following the treatment found in The Sibley Guide to Birds, *which lists closely related species near one another and the most highly derived species toward the end of the sequence.*

- ☐ Ovenbird
- ☐ Worm-eating Warbler
- ☐ Louisiana Waterthrush
- ☐ Northern Waterthrush
- ☐ Golden-winged Warbler
- ☐ Blue-winged Warbler
- ☐ Black-and-white Warbler
- ☐ Prothonotary Warbler
- ☐ Swainson's Warbler
- ☐ Tennessee Warbler
- ☐ Orange-crowned Warbler
- ☐ Nashville Warbler
- ☐ Connecticut Warbler
- ☐ Mourning Warbler
- ☐ Kentucky Warbler
- ☐ Common Yellowthroat
- ☐ Hooded Warbler
- ☐ American Redstart
- ☐ Kirtland's Warbler

- ☐ Cape May Warbler
- ☐ Cerulean Warbler
- ☐ Northern Parula
- ☐ Magnolia Warbler
- ☐ Bay-breasted Warbler
- ☐ Blackburnian Warbler
- ☐ Yellow Warbler
- ☐ Chestnut-sided Warbler
- ☐ Blackpoll Warbler
- ☐ Black-throated Blue Warbler
- ☐ Palm Warbler
- ☐ Pine Warbler
- ☐ Myrtle Warbler
- ☐ Yellow-throated Warbler
- ☐ Prairie Warbler
- ☐ Black-throated Green Warbler
- ☐ Canada Warbler
- ☐ Wilson's Warbler

INTRODUCTION

An interest in nature leads you into a kind of enchanted labyrinth.
You wander from corridor to corridor; one interest leads to another
interest; one discovery to another discovery. It matters little where
you begin.

—EDWIN WAY TEALE, *Circle of the Seasons*

n February 1947, the naturalist Edwin Way Teale and his wife, Nellie, took an adventurous automobile trip in search of nature. Driving on country roads through farmland and prairie, forest and marsh, mountains and swamp, the Teales tracked the progress of spring from the Florida Everglades to northern New England. Moving north as the days lengthened and warmed, songbirds of many species greeted the Teales through the American landscape. The Teales took in local history and culture and explored little-known green spaces as they followed the birds' own journeys. The account of their travels, *North with the Spring*, the first of Edwin Way Teale's grand quartet of seasonal natural-history travelogues, was published in 1951.

I first encountered Teale's tale of spring migration in the late '50s, when my mother read passages from it to me and my brother on our cozy living room sofa before bedtime. Our mother, a single parent who worked as a dental assistant to pay the bills, was rediscovering her own childhood love of nature through her avocational reading of Teale, Henry David Thoreau, Isaak Dinesen, Hal Borland, Elspeth Huxley, and others, prowling the local library in search of nature in the written word and bringing her discoveries home to share with us. Thus, long before I'd even decided to become an ornithologist, in the days when I was first exploring the nearby woods and fields, Teale's migrant birds became nighttime visitors to our home through our mother's reading.

At that time, the northern neighborhoods of Baltimore, where I grew up, were still dotted with forest tracts. Wood Thrushes sang from the shade of the woodsy stream bottoms, and coveys of Bobwhite Quail visited our backyard garden. My brother, Bill, and I were free to range widely for hours. When not in school, we children wandered, discovering the outdoors on our hikes to Tramp Rock in nearby Gilman Woods. Midway through spring, in their murmuring flocks, Myrtle Warblers flashed their golden rumps in Red Maples that were just starting to push out their pale pinkish-green leaves. Bill and I would take note of these northbound migrants as well as the monotonous but cheerful territorial singing of Northern Flickers. At the end of the

---

OPPOSITE: Blue-winged Warbler

day, we were called home by the sound of the brass school bell that our mother rang to signal us in for dinner.

In 2015, after a lifetime spent observing birds around the world, I decided to take a road trip up the Mississippi Valley to the Great North Woods to experience spring migration as the Teales had done, but along a different pathway. I would first meet the migrant birds on the coast of Texas after they had made their journey across the Gulf of Mexico, returning stateside from their tropical wintering grounds. Then, tracking spring's march up through the continent, I would follow the migrants as they moved northward in fits and starts. I would meet these birds at an array of favored stopover sites and midcountry nesting spots from Texas to Canada, and I'd seek out the most peripatetic of the migrants on their nesting territories in the boreal forests of northern Ontario and high in the Adirondack Mountains of northern New York State. I would travel the backroads and camp in the woods. Each night, from the comfort of my tent, I would listen to the last of the evening's birdsong, and each morning I'd wake to their cheerful dawn voices.

My trip had several purposes beyond harkening back to the Teales' own journey. As a practicing field naturalist, I wanted to experience, first-hand, spring migration at its most effulgent—right through the country's midsection. I wanted to witness spring's procession up the great riverway and spend time in the prettiest patches of green scattered up and down the Mississippi. As a research biologist, I also hoped to learn about the people who are studying migratory birds in the field. And, having experienced first-hand the long-term declines of migratory birds, I wanted to see the work that various institutions are doing to restore migration to its former abundance.

I planned to focus my hundred-day birding excursion on the wood warblers. These brightly patterned songbirds constitute the heart of the continent's avian migratory system, and their well-being is a bellwether for the health of other bird species and of the land itself. Thirty-seven species of wood warblers breed in eastern North America, and I wanted to search out and observe each in its breeding habitat.

But could I find *all* of them on their nesting territories in a single field trip? Whether or not I could, I knew that my search would lead me to the finest natural places along my transect, and that on my hunt I would encounter many other fascinating birds and all sorts of wildlife and surprising natural phenomena.

At the end of the northern leg of my long circuit lay Ontario, that great boreal nesting ground of the wood warblers. I'd wondered about it for decades, wanting to spend time there in late spring, when the male warblers are singing loudly and establishing territories. But northern Ontario was something of a natural-history black hole. It is still undeveloped country, lightly settled Ojibway First Nations traditional lands, with Black Bears and Moose, Timber Wolves and Wolverines. I had a hunch that the region—north-country wilderness rich in big game—might produce an extravaganza of wood warblers, so I planned to drive up to the very end of the Nord Road and set up camp at the height of songbird breeding season.

As a coda, meant to complete my thirty-seven-species warbler checklist, I planned to end my field trip in the Adirondacks, where I would climb a mountain I had last climbed forty years ago. The last quest bird on the list, the Blackpoll Warbler, would await me on that mountaintop. I had spent my childhood summers hiking in the Adirondacks, a vast forest landscape that left an indelible impression on me; I even wrote a book about Adirondack birds that I published in 1978. Yet it had been nearly half a century since I had spent serious time there, and so to complete my springtime quest I would climb the mountain that I'd first climbed in 1967, and camp in a place where I'd last camped in 1977. I would see first-hand how the park's forests and birdlife had fared since I had last worked there. What surprises—good or bad—would I find upon my return?

But the true motivation for my trip was, of course, the fact that I am and remain a lifelong birdwatcher. From my boyhood years in Baltimore and my summers spent in the Adirondacks to my working years spent in the rainforests of New Guinea, I have studied birds both professionally and avocationally for more than half a century. Over the years, watching birds on spring mornings has created enduring memories of the annual passage of migration.

On May 6, 1967, I stood in the creek-bottom forest near my grandparents' home, with a pair of heavy 7x50 binoculars that my grandfather had loaned me. It was a Saturday morning—the day of Maryland's annual May Count of birds—and I was surveying birds in this north Baltimore forest remnant. My keen fifteen-year-old ears picked up an unrecognized birdsong in the canopy vegetation: *bizz-buzz*. What was it? I focused the clunky field glasses on the singer, a small yellow bird with blue-gray wings and white wing-bars, perched on a twig about thirty feet up. Plate 50 of Roger Tory Peterson's *Field Guide to the Birds* confirmed it: this was a Blue-winged Warbler, and it was Life-Bird number 230 for me. Checking my field guide was simply pro forma; like any young birder, I voraciously consumed ornithological reference books and had learned the local birds cold before I'd seen even a small fraction of them. My passion was to see in the wild the birds that I had learned by heart in evenings at home with my small collection of reference books. Birdwatching came to overshadow my other natural-history interests—minerals, fossils, and butterflies.

I recently leafed through the diary in which I'd documented that sighting of the Blue-winged Warbler. I have seen scores of Blue-winged Warblers since that day, but my first view of the little singing bird will remain forever burned in my memory. Indeed, the chief reward of springtime birdwatching is the well-etched memories of the sights and sounds of the species that migrate north each spring and south each fall. (By 2015, the time of my big field trip, I was having difficulty hearing the song of the Blue-winged Warbler as well as those of some other species that give very high-pitched vocalizations.)

The late 1950s and early 1960s, the decades of my childhood, witnessed a remarkable flowering of nature appreciation and environmentalism in the United States, prompted by *Audubon* magazine,

Peterson's field guides, and the work of an array of popular nature writers who included Teale, Hal Borland, Peterson, Rachel Carson, Louis Halle, and others. Local Audubon societies sponsored nature walks and hosted evening lectures by national luminaries, including Peterson and Teale. This spring tide of natural-history appreciation culminated with the first celebration of Earth Day, in April 1970. Through it all, I never forgot Teale's vivid recounting of the waves of wood warblers sweeping northward through the regenerating upland forests of the Appalachian Mountains. He wrote of places I longed to visit and sights I dreamed of seeing. He planted seeds that lay dormant for decades.

College provided my escape from the parochial confines of life in an average East Coast town, and a postgraduate travel fellowship took me to New Guinea for fifteen months, where I practiced field ornithology without anyone looking over my shoulder. Because of that stint, I won a fellowship for a doctoral program in biology at Princeton, and after two years of coursework I returned to New Guinea for twenty-nine months spent studying birds of paradise in a patch of grand old-growth rainforest at an elevation where the climate year-round is reminiscent of mid-April in Washington, D.C. During my professional years, I mixed ornithological field studies with ecological research and nature conservation, working in a number of countries in tropical Asia and the Pacific, far from my Mid-Atlantic boyhood haunts. My mother was fascinated by my work but couldn't completely grasp its details or its faraway locales. Instead she focused her love of nature close to home. She remained in Baltimore and continued birdwatching locally, while I traveled the world on ornithological and conservation-related pursuits. Yet her steadfast love of nature *near home* reminded me of its importance; the wonder of spring migration on the East Coast kept me loyal.

Fast-forward to May 11, 1995. I joined birding buddies Peter Kaestner, Chuck Burg, and John Anderton to crisscross the countryside northwest of D.C. in our first attempt at a "Warbler Big Day"—we were seeking as many species of migrant wood warblers as possible in a single frantic day, done not for some statewide census but strictly for fun. Starting well before dawn (warblers are at their most vocal and visible early), we visited forests along the Patuxent River, Little Bennett Regional Park, the Chesapeake & Ohio Canal, the Potomac River, and Sugarloaf Mountain in a race against time. We were lucky that the whole day was cool, misty, and damp, which encouraged the warblers to stay active and visible to our little party seeking to push the limits of the possible. That day we recorded five Blue-winged Warblers, plus 169 other individual warblers, totaling *thirty-one* different species, all of them migrants arriving from their various southern wintering areas in a blaze of colorful spring plumages and exuberant male territorial song. Recording more than thirty species of warblers in a day is a difficult feat, so we were elated by our accomplishment, which we have repeated only a few times since. For many of us, these annual Warbler Big Days have provided some of the most brilliant memories of our birding lives. Such days combine lovely verdant habitats, wildflowers, colorful birds, other wildlife, and fine companionship—the sort of experience I hoped to recapture on my big birding trip up the Mississippi.

The death of my mother in the spring of 2013 was another incentive for my journey. Her passing rekindled bittersweet recollections of childhood and my first encounters with nature: a Snout Butterfly! A fossilized Miocene-era crocodile tooth! A Pink Lady's Slipper orchid! A singing Baltimore Oriole! Her death awakened, too, my memories of those evening readings of Teale and the other great naturalists. My thoughts returned to the Teales' epic journey tracking spring's passage northward to the Canadian border. For thirty-five years, I had worked in lands far from home and was able to pursue North American birds as a hobby for only a few days each spring. Now my approaching retirement promised freedom, and a partnership with the American Bird Conservancy allowed me to spend the entire spring of 2015 investigating America's bird-migration spectacle.

My desire was not to replicate the Teales' journey but to recraft it in a manner relevant to my own interests in Neotropical migratory songbirds and the conservation challenges they are facing in the twenty-first century. Teale had focused on the broad phenomenon of spring. My focus would be on the spring migration of birds. Teale had begun in Florida and pursued a route that trended southwest to northeast, but I believed that a route following the main stem of the Mississippi was more promising. Teale's route ran the length of the Appalachians, but my goal was to follow the movement of the bulk of the songbirds coming from the Tropics. After I'd decided on my course, I worked with the American Bird Conservancy and the Smithsonian Institution, plus dozens of local specialists, to choose the places I should visit to find nature at its best. By the time I set down my maps, I'd charted a route that would take me through nineteen states plus a broad swath of Canada.

From the start, I'd planned to document my journey as Teale had done, taking readers along with me as I traveled with the spring songbird migration, painting a picture of life on the road and the sights and sounds of the bird species and the little-known places that I'd visit. And I planned to share what I learned from the people and institutions working to study, conserve, and restore America's natural treasures. As a result, my story does not confine itself to migrant songbirds; it includes all sorts of birds, other wildlife, local history, foods, and stories, so that readers can learn, as the Teales did, the very flavor of the lands where I traveled and camped and birdwatched. Thus this is a book about far more than songbirds or a long-distance natural history road trip. It's about the birdwatcher's quest itself, and the scientific curiosity that it epitomizes, as I experienced it in a hundred days on the road with the birds.

ONE

Birds of Spring
March 29, 2015

Wood warblers come north as the leaves unfold. They feed on the forest caterpillars that feed on the new green leaves. Their northward flight keeps pace with the unfolding bud and expanding leaf. . . . Buds burst, new leaves unfurl, larvae hatch, and warblers appear.

—EDWIN WAY TEALE, *North with the Spring*

n the eastern and central United States during the spring songbird migration, new colors and sounds flood northward across the landscape. Although generally overlooked by most people, the songbirds travel through woods, cities, and suburban neighborhoods for a few weeks in large numbers, showing their bright breeding plumages in backyards as the sap is rising, the leaves are flushing out, and the woodland wildflowers are blooming. Everything is fresh and green, and the migrant birds are vibrant window dressing to this most seductive of nature's seasons.

We birders love the sounds of spring song, the visitors' resplendent plumages, and the fecund, damp-loam aroma of the new season in the woods. Males of many songbird species sing their territorial songs and forage greedily to refuel for their next move northward. Birding most months of the year is pleasant enough, but birding on certain days of spring is transporting. Within a half-hour's drive of my home in Bethesda, Maryland, I can track down nearly a hundred species of birds—including more than twenty species of wood warblers, five vireos, five flycatchers, five thrushes, two tanagers, two orioles, and others—on a single good day in May.

THE NATURAL HISTORY OF MIGRATION

The birds we glimpse passing northward in spring are conducting a strategic relocation exercise to reproduce and to enhance the long-term survival of their genes. The birds cross seas and continents in their travels in spring, just as they do in autumn, to situate themselves in just the right place at the right season. Migration, then, is the purposeful and directed seasonal movement from one geography to another in a repeated annual cycle—a life-history solution to the ecological challenges posed by living in a seasonally variable world.

Many animals migrate: butterflies, dragonflies, salamanders, birds, bats, and whales, to name a few. Migration is most prevalent in environments that exhibit predictable changes in environmental conditions with the seasons: typically from wet to dry or from warm

OPPOSITE: Blackpoll Warbler

to cold. Animals migrate to escape from the cold, to find food, and, in some instances, to reach places where they can reproduce safely. Think of female Gray Whales, traveling from Alaska to the warm and secluded waters of the Gulf of California each year to give birth. Some migrations are grand, such as the seasonal movement of wildebeest across the Serengeti. Other migrations are less substantial, as when our local populations of American Robins join into flocks and wander within the Mid-Atlantic region in search of winter fruit after the ground has frozen and they can no longer find earthworms. Some migrations are just a matter of a few hundred meters, as when Spotted Salamanders move from their forest floor home ranges to vernal pools to reproduce. Migratory movements come in so many forms that whole books have been written to describe the breadth of the phenomenon.

In the Mid-Atlantic, some birds migrate and some do not, and the new birdwatcher might wonder why. Those that do not migrate seem to have the capacity to provision themselves on their home range year-round, even during the leanest months of winter. Those that cannot find adequate food during the winter move to another location where they can. What advantages accrue to species that do *not* migrate? To answer that, first let's acknowledge that in nature, long-distance travel is dangerous. The migrant might fall prey to predators, get lost, or have difficulty finding food along the way. The migrant also might have trouble finding an ideal wintering environment that is not already occupied by competitors, or it might encounter bad weather en route. Staying at home, by contrast, has advantages. The resident bird is familiar with the landscape, is aware of local threats, and knows where to find food. Typical nonmigratory birds have a broad and seasonally adapting diet that allows them to meet the challenges of winter. An overwintering chickadee, for example, might visit the same backyard feeder each winter. It probably inhabits a territorial patch of no more than fifty acres over a full year. By staying put, the bird has a reliable habitat in which to feed and reproduce. It gets to know the nooks and crannies of its own patch and, year by year, knowledge of its

North on the Wing

home range grows, and accumulated information about local threats helps it to avoid them. Staying at home is the best bet for certain species.

What about advantages accruing to the migratory bird? First, let's remind ourselves that the spring migration of birds is ultimately motivated by the physics of the earth's orbit around the sun. Spring happens because the earth's axis of rotation is tilted 23.5 degrees away from vertical with respect to its annual orbit of the sun. As the earth circles the sun, its poles gradually tilt toward and then away from the sun. As the earth moves from its northern winter solstice to spring equinox, more direct sunlight hits the earth's surface in the Northern Hemisphere and day length increases. These two physical changes to our north-temperate sector of the planet produce the procession of spring. Longer days and increased daily solar radiation encourage the rapid growth of plant life and the invertebrates that subsist on plants. Birds move north to follow this burst of plant and insect life as the sun moves northward with the season. The migrants time their travel to arrive at their breeding habitat at a point of maximum seasonal abundance.

Tropical forest, where many songbird species spend the winter, is benign but lacks the very long days of northern summer that give birds plenty of time to forage, and it also lacks the seasonal flush of biotic growth that accompanies springtime in northern climes. During the nesting season, most migrant songbirds eat mainly insects, and thus their annual movements are partly tied to tracking the abundance and availability of insect prey. Insects are abundant in boreal forests in late spring and summer but essentially absent during the cold and dark time of the year, from late autumn to early spring, so the songbirds migrate to the Tropics to escape the long season of privation. In both autumn and spring, the signal to migrate comes from the sun—day length and the sun's track across the sky.

In spring, most migrant songbirds make an effort to return to their natal territory, settling quite close to where they were raised. In addition, in their second autumn, they return south to the same area where they successfully wintered the preceding year. How they achieve the GPS-like adjustments to locate these pinpoints on the globe remains a mystery, as we will discuss at length later in the book, but we do know that many birds accomplish this remarkable feat of orientation year after year.

SONGBIRDS, NEOTROPICAL MIGRATORY BIRDS, AND WARBLERS

As the days lengthen in the United States, most of the small land birds that migrate here from south of the border in springtime are songbirds. Even non-birdwatchers know plenty of these creatures; examples include the Northern Mockingbird, Blue Jay, Carolina Chickadee, Yellow Warbler, Northern Cardinal, House Wren, and American Crow. Their lineage—the suborder Passeri within the order Passeriformes (the perching birds)—globally comprises about four thousand species of land birds that evolved in the Australian region fifty million years ago, when it was part of Gondwanaland. Today songbirds inhabit every continent except Antarctica. As their name suggests, they are accomplished singers: they alone among the birds possess a highly evolved syrinx, a complex vocal organ that allows them to make their elaborate and distinctive sounds. That said, most birds are regular vocalists, but those without this anatomic specialization cannot match the singing capabilities of the songbirds, which make memorable and affecting music.

Most familiar Neotropical migrants are songbirds, but there are exceptions. One group, the tyrant flycatchers (such as the Eastern Kingbird and Great Crested Flycatcher), are passerines but not songbirds. For simplicity in this book, however, I use the term *migrant songbird* as a convenient shorthand to refer to all migrant perching birds, including the flycatchers.

In this book, we shall encounter many kinds of Neotropical migratory birds, meaning those that breed in Canada and the United States and migrate south in autumn to spend the nonbreeding winter season in Mexico, Central America, South America, and the Caribbean. They include more than two hundred species of land birds, plus many other groups: waterfowl, raptors, shorebirds, and terns. Some of these species, such as the Red Knot, travel as much as ten thousand miles on their southbound journey in fall and do it again on their return in spring. All these birds are border-crossing international travelers.

Special among the Neotropical migrants are the New World wood warblers (the Parulidae), a unique Western Hemisphere family of 118 species of small songbirds. As a family, they are widespread and well known, but individual species often are tantalizingly difficult to locate on their breeding grounds. Forty-seven of the wood warbler species nest in the United States and Canada and migrate to winter in the Subtropics and Tropics. Warblers that nest north of the Mexican border are the passionate focus of many birders, especially during spring migration; among these are the thirty-seven eastern wood warbler species that I hoped to see on their breeding territories on the journey described in this book. Part of their allure is their elusiveness—how many of us have seen all thirty-seven in a spring season? How many of us have seen all of them on their breeding grounds?

THE TENNESSEE WARBLER

Often in this book, I will highlight a particular wood warbler when describing an aspect of the life history of the Neotropical migrant songbirds. The Tennessee Warbler, for example, displays the migratory pattern and life cycle of a prototypical Neotropical migratory songbird. This bird—a four-inch-long warbler wearing a subtle plumage of gray, olive green, and white—winters in southern Central America and northern South America and breeds in the boreal forests of the northernmost United States and Canada. In late June, the Tennessee Warbler hatches from its egg in a small cup nest hidden amid low vegetation on the ground in a northern forest clearing. It starts its life in a time of long days, warm temperatures, and abundant insect prey. It is provisioned by its two parents while in the nest, which it departs after only eleven days. Its parents continue to bring it meals of small insects as it learns to fly and begins to move about the environs of its local forest patch with its siblings. By mid-August, it can fly well and is

Tennessee Warbler

able to forage for itself—quite an accomplishment for a bird less than two months old.

In late August, feeling the tug of the innate impulse that scientists call migratory restlessness, our Tennessee Warbler takes to the sky after dark, heading southeastward in a series of night flights. Each flight might last as long as ten hours and cover more than two hundred miles, punctuated by stops to rest and refuel. After several weeks, it reaches the Gulf states. There it halts to build up stores of fat for its overwater trek to the Tropics. One evening after dark, it heads out over the Gulf to Cuba or another Caribbean stopover. After a few days of refueling, the bird once again takes off after nightfall and heads to northern South America or southern Central America, where it will spend the winter. In its chosen winter home, it seeks out a large patch of habitat to pass the next seven months of its sociable life, joining other Tennessee Warblers as well as other warbler species in flocks that forage for food each day. This flocking behavior probably helps the birds to find food more efficiently and avoid predators.

With the lengthening of the days in March, the warbler, in its tropical retreat, becomes restless once again. In early April, on a clear, warm evening with light southerly winds, it departs its forest patch, climbs into the air, and heads northward in stages to its breeding habitat in the Great North Woods. Its main challenges are crossing the Caribbean and then the Gulf of Mexico, which it presumably achieves in separate night flights.

Back in the United States, it rests, refuels, and starts a series of nocturnal flights northward up the Mississippi Valley before it faces one more physical hurdle: the Great Lakes. We know that songbirds consider crossing the lakes a substantial challenge because of evidence from Magee Marsh, an important bird reserve on the south shore of Lake Erie. Under ideal springtime conditions, large numbers of north-bound Tennessee Warblers settle into the woods here on the south side of the lake rather than continuing on across the lake. Instead, they make the lake crossing after they have refueled and while they are fresh at the start of a new flight northward.

Once back in the Great North Woods, each migrant returns to the boreal forest where it was born. The males arrive about ten days before the females because they must win a breeding territory to attract a mate once the females arrive on the breeding grounds. To do this, a male must defend his territory from competing males. His song helps him: one of its functions is to declare ownership of his territory, while the other is to attract a mate. The male bird sings his loud and staccato series of chip notes more than a thousand times a day. Once mated, the Tennessee Warbler's life cycle begins all over again.

The most important measure of any organism's life is reproduction. For a warbler, this takes place during a relatively brief period in June and July. Success in lifetime reproductive output among individuals drives natural selection, which is the basis for organic evolution. We might call the reproductive drive the "life force." For human societies, wealth, knowledge, and culture constitute major life goals, but for species living as a part of wild nature, the single objective is to leave on earth an abundance of fecund offspring, thereby ensuring the survival of their genes in following generations. Thus, when a male Tennessee Warbler gives its staccato song, it is not singing because it is happy (though it may be) but because it is trying to do three important things: stake a claim on a patch of territory, signal its presence to other, competing males, and attract a mate from among the females in the vicinity. Male migrant songbirds also sing during migration—less commonly far from their breeding range, but more commonly as they get close to their nesting grounds. It appears they are "practicing" their territorial song, so they can put it to good use immediately upon their arrival.

At the end of summer, adult Tennessee Warblers and their young migrate southward separately and probably follow different routes and timetables. The first-year birds get together and head south for the first time with nothing more than genetically programmed rules to guide them toward their tropical winter destination, where they have never been. That many of these yearlings arrive safely at their destination is testament to the fundamental soundness of their evolved migration system, as well as the mysterious nature of the phenomenon. To complete the roundtrip journey south and then back north to the breeding area, the warbler needs, in essence, a map, a compass, a calendar, a clock, and a good memory, all stashed away in a brain little bigger than a couple of peas. It is no wonder that biologists look upon migration as a miracle of evolution. That said, many individual warblers perish during the two legs of migration each year. Typically, Tennessee Warblers that successfully fledge from the nest live only three to four years, and only half the birds that head south return to their northern birthplace the following spring.

So what advantages accrue to the Tennessee Warbler that impel its risky twice-yearly migration to and from the Tropics? Key among them is food, of course. Seasonal dietary specialization by the warbler leads it to follow its preferred arthropod prey from tropical to northern temperate forests each year. In the northern summer, this warbler raises a brood of four or more nestlings in a boreal forest where the long days produce a prodigious flush of insect prey, which the warbler adults harvest to feed to their nestlings. Later, as the northern winter approaches, migrating allows them to find a habitat in the Tropics that provides insects as well as nectar sources.

Between September and April—for as long as eight months a year—the typical Tennessee Warbler resides in some patch of habitat in the Tropics. One might assume this is a time of relaxation for the bird, but that's probably not so; instead it is a time of survival. This is evidenced by the fact that many of these wintering birds have not added fat stores by the end of their sojourn in the Tropics, indicating that they haven't found abundant food there. This is the nonproductive season, when the bird is biding its time. But it is an important period nonetheless, for the bird must remain healthy and fit during this off-season if it is to succeed during its hectic breeding season in the North Woods.

THE CONSERVATION STORY

The life story of any migratory bird such as the Tennessee Warbler is incomplete without an account of the threats that these birds face during migration. Various human-caused threats were noted in the 1960s by Rachel Carson, Roger Tory Peterson, and other prescient naturalists. More than half a century later, we have measured the specific impacts of those threats. Radar-based studies of migration across the Gulf, as well as long-term studies of occupancy by breeding wood warblers in forests in New England and on their tropical wintering grounds, indicate that the populations of Neotropical songbirds have decreased by more than half over the past fifty years.

An array of forces, operating in different portions of the birds' ranges, has brought about this serious decline. Loss of habitat has occurred mainly on wintering grounds and along migratory pathways, while fragmentation of breeding habitat has led to reduced nesting success for some species because of the increasing abundance of lesser predators such as opossums and raccoons and because of nest parasitism by the Brown-headed Cowbird. Cats, lighted towers and buildings, glass windows, and wind farms also take a mortal toll. And, without doubt, climate change has had an impact, disconnecting the timing of the birds' movements from the leafing-out of canopy vegetation and the attendant peaking of populations of the leaf-eating insects that are the birds' primary prey.

Various groups and institutions, both large and small, are battling the threats that have led to the decline of migratory birds. In addition, research teams at scores of colleges and universities are studying these impacts and seeking solutions, while federal, state, and local governments work to preserve and restore strategically important natural habitats. Many nongovernmental organizations aid the effort as well. Every group involved understands the need to address the threats that birds face at different stages of their life cycle, ensuring that no vulnerability in the bird's life is overlooked.

Because conservation is an essential facet of the spring migration story, it was one of my primary interests during my spring songbird quest. I visited and spoke with many people who work on programs to conserve migratory birds and the migratory system as a whole. They are the heroes of bird conservation in North America, and my reports on their work, scattered throughout the book, share both the lessons they have learned and the hope for the migrants' future that their stories have given me.

As I planned my journey, I knew I would travel in very different circumstances than the Teales did on their 1947 journey. While they had stayed in roadside motels, I camped in a tent every night. This not only saved a bundle of money but also kept me connected to the birds. For a hundred days and nights, I was surrounded by the sights and sounds of nature. I carried a two-burner gas stove and a cooler stocked with food, only occasionally eating in restaurants and instead dining al fresco at my campsite. The Teales had traveled through a world without interstate highways, Walmarts, Japanese SUVs, or East Coast suburban sprawl. I experienced phenomena unknown to the Teales as I traveled through a world of declining rural towns in the Deep South and Rust Belt cities in the North, bringing along my high-tech hand-held devices. But we were both in search of the same thing: spring's annual march through wild America.

The Teales had traveled in a black Buick without seatbelts, stereo, or GPS. I drove a second-hand Nissan SUV carrying a kayak and a bike. The car got me to desirable campsites in various patches of woods, from which my bike, kayak, and foot power took me to interesting natural locations nearby. The SUV held storage containers stocked with maps, field guides, and local travel guides, as well as classic accounts of similar trips—*North with the Spring*, of course, as well as *Wild America* and *On the Road with John James Audubon*. Duffle bags held sleeping pads and bags, pillows, field clothing, and rain gear. I had digital camera equipment and two cell phones, a portable Wi-Fi hotspot, a GoPro video camera, and a laptop computer, which I used to record the journey and upload twice-weekly posts to a travel blog on the American Bird Conservancy website so people could follow the trip online. As I prepared, it became clear that even though I'd camp out and make meals in circumstances more rustic than those the Teales encountered in 1947, mine would be an utterly twenty-first-century operation.

Now let's take to the road. Let's get away from the cities and head south. Let's fill the ice chest with food and drink, get onto Interstate 81, set the cruise control, and make our way to Texas.

TWO

The Texas Gulf Coast

FIRST LANDFALL

Early April 2015

To the green mist of the cypresses and the moving clouds of the swallows we could add the movement of the stars as a sign of the sure approach of the spring. With all the galaxies and planets and stars, the solar system was setting the stage.

—EDWIN WAY TEALE, *North with the Spring*

The first port of call for many trans-Gulf migrant songbirds in spring is the coast of southeastern Texas, and that is where I plan to meet them. At the end of March, I depart Maryland and over three long days drive to Mad Island, on the Texas coast between Freeport and Corpus Christi. About sixty miles south of Houston, Mad Island is where I'll meet the vanguard of the northbound migrants. My declared field trip focus is songbirds, but spring on the coast of southeastern Texas is the height of distraction for the curious naturalist because of its abundance of waterbirds, wildflowers, snakes, lizards, and more. At Mad Island, in fact, I'll learn more about coastal prairie than I will about songbirds, which have been slow in arriving stateside this particular spring. Every turn of my trip, as I will learn, will introduce me to both little-known ecosystems and the people working to conserve these habitats and the diverse bird communities that depend upon them. I have come for songbirds, but I'll encounter much more.

Mad Island is not, despite its name, an island at all but a tract of prairie overlooking Matagorda Bay and protected from the Gulf by the long, narrow barrier island that runs nearly unbroken southwestward to the border with Mexico. The Clive Runnells Family Mad Island Marsh Preserve, a seven-thousand-acre tract of remnant coastal prairie under the management of the Texas chapter of the Nature Conservancy, lies just southwest of the hamlet of Wadsworth. This preserve, the southernmost point on my circuit, was the official starting point of my journey north to Canada. Here I visited a bird-banding project of the Smithsonian's Migratory Bird Center (SMBC). Directed by Peter Marra, a Smithsonian colleague, the SMBC studies the ecology and conservation biology of migratory birds in the Western Hemisphere. A SMBC field team seasonally stations itself at the preserve, which generously allows the researchers to do their work here. Marra had advised me to visit the bird-banding projects at Mad Island and at Grand Chenier, Louisiana, to discover what's being done on the ground to investigate the spring migration of songbirds across the Gulf of Mexico.

OPPOSITE: Common Yellowthroat

With the Colorado River estuary to the east, Matagorda Bay to the south, and Tres Palacios Bay to the west, the Nature Conservancy reserve at Mad Island is a bird heaven year-round. During one recent winter, in fact, the wetlands and rice fields of the Mad Island region supported more than two hundred thousand waterfowl, and Mad Island is known for periodically generating the largest species list of any regional Christmas Bird Count (CBC) in America. The annual CBC, sponsored by the National Audubon Society, is the nation's longest-running citizen-led annual birding event. Groups of volunteer birdwatchers all over the country count the birds they see and hear within a designated "count circle" fifteen miles in diameter on a designated day between mid-December and early January; more than two thousand such counts are conducted across the continent each year. CBC records on the changing status of winter bird populations have been compiled into a century's worth of field data for researchers. Whereas the CBC censuses wintering bird populations, my arrival at Mad Island was timed to match the return flight north of many birds that wintered to the south. In preparation for the coming returnees, I set up my tent on the edge of a broad expanse of coastal prairie, just a short drive down a sandy track from the banding station at the coast.

MIGRANT WINTER LIFE IN THE TROPICS

When I showed up, the bulk of Neotropical migrants that the Mad Island banding crew awaited were still down in the Tropics, and I wondered whether the evolutionary origins of these migratory species were northern or southern. Although the details are messy, recent evolutionary analyses indicate that most Neotropical migrant lineages evolved in the temperate zone and subsequently adapted to spend the winter in the Tropics. Wood warblers are a mixed bag: most sublineages of the group evolved in North America, not the Tropics. So some birds evolved in the temperate zone and shifted their wintering habitat south, while others evolved in the Tropics and shifted their breeding habitat north, as originally sedentary lineages began to migrate to follow the seasonal availability of resources. The most likely factor motivating their transition to the migratory habit was the cycle of glacial advances and retreats over the past 2.5 million years. Glaciation's major impact was the enforced relocation of breeding ranges for virtually all the North American songbirds, and it also had major influences on rainfall and seasonal drought in the Tropics.

What is life in the Tropics like for Neotropical migrants? The migrants have to share their tropical habitat with the abundant local resident birdlife, and they are in this habitat during the dry season, when resource levels are depressed. The migrants likely struggle to establish a winter foraging space, and they must work hard to prepare their bodies for the return flight north and the demands of nesting and raising offspring.

Many migrants seem to select a winter habitat in the Tropics that comes as close as possible to their habitat in the north temperate and boreal zones of North America, but there are limits to their ability to match habitats between the temperate and tropical zones. The tropical flora is different, the days are shorter, the weather typically is warmer, and the food resources differ, offering substantially more fruit and nectar than the northern habitat does. The differences probably outweigh any similarities the birds are likely to find, and accordingly some species change their feeding behavior between summer and winter—those that are specialized insect eaters in the north shift, in some instances, to consuming nectar or small fruits in the Tropics.

Migrants' social behavior also can differ between north and south. Most Neotropical migrants establish breeding territories in the north, with a male-female pair occupying each territory and raising their young there. In the Tropics, by contrast, some species establish solitary (one-bird) territories for much of the winter, whereas others establish no territory at all, instead roaming about and joining other foraging species of migrants and nonmigrants. Some migrants, for example, take up with multispecies flocks of birds that regularly follow army ants, pursuing the invertebrates that flee before the moving ant columns. Other species become seasonally sociable with their own kind, foraging in monospecific flocks on flowering and nectar resources. And some species are reported to roost in single-species groups at night.

Those that join mixed-species foraging flocks in the Tropics are part of a phenomenon that seems widespread in the Tropics of both the New and the Old World, one that may be driven, at least in part, by the threat of predation by bird-eating raptors, snakes, and mammals. The mixed flock occupies a large foraging territory each day, and as the group moves about, individual birds or bird pairs join up or drop out of the flock as it passes through their home ranges. A wintering thrush might forage alone on its solitary home range but join a flock to forage with the group as it passes.

Studies of Kirtland's Warbler have demonstrated that winter conditions influence subsequent breeding productivity. Sarah Rockwell and her collaborators, for example, have shown that ample rainfall in March on the Kirtland's wintering ground in the Bahamas led to both earlier arrival by males back in Michigan and the production of more fledglings per male. So winter conditions for a warbler in one location can have measurable carryover effects on breeding results in a distant locale. Similar results have been shown for American Redstarts that winter in the Caribbean and breed in eastern North America.

During the long and lean season in the Tropics, each migrant songbird has two clear objectives: to avoid being eaten by a predator and to maintain (or improve) its physical condition in readiness for the migration north and the breeding season to come.

THE COASTAL PRAIRIE ECOSYSTEM

Mad Island lies at the heart of the coastal prairie ecosystem that once dominated southwestern Louisiana and coastal Texas. The ecosystem, decimated by development, once included extensive tallgrass prairie, wetlands, and patches of gallery forest. Today, it is one of the most endangered natural habitats in all of North America. This narrow band of prairie lies just back from the coastal marshes and once stretched in an arc that paralleled the coast of the Gulf of Mexico. But less than 1 percent remains in its original state. Most has been degraded by cattle grazing, rice farming, sugar cane monoculture, oil and gas development, and creeping urbanization. Public and private restoration of prairie lands is crucial to the future of this critically endangered ecosystem, which is home to an array of birds—including the Mottled Duck, Attwater's Prairie-Chicken, White-tailed Hawk, Crested Caracara, Scissor-tailed Flycatcher, Painted Bunting, and Dickcissel. Prior to western settlement, this unique tallgrass prairie evolved through the influences of seasonal rainfall, periodic fire, and grazing by Bison and other wild ungulates; native grasses thrive under conditions in which they can outcompete woody plants. In the intact areas of the coastal prairie, native grasses such as Big Bluestem, Indiangrass, Eastern Gamagrass, and many native wildflowers dominate. These species cannot tolerate heavy year-round grazing by domestic cattle, which encourages the invasion of exotic annual grasses such as Vasey Grass, from South America, and Johnson Grass, from the Mediterranean. Natural prairie is dominated by long-lived perennials, and, with a few exceptions, annuals are rare in undisturbed prairie sod.

Most Neotropical migrant songbirds that cross the United States–Mexico border in spring have destinations far north of Texas. That said, a few do settle to breed in the Texas coastal prairie, including the Scissor-tailed Flycatcher, Common Yellowthroat, Painted Bunting, and Dickcissel. The first of these, with its long swallow tails and apricot underwings, is among the loveliest birds in America. It winters in open country in southern Mexico and Central America and breeds in the southern Great Plains all the way down to Mad Island, where it is one of the most familiar birds along rural roadsides. The Common Yellowthroat—one of my quest group, the eastern wood warblers—winters as far south as Panama. The only wood warbler to breed right along the Gulf Coast, this diminutive, bandit-masked songster is North America's most widespread wood warbler. The yellowthroat is unusual in breeding in marshes and shrublands, places that often lack woods or trees. Most breeding wood warblers prefer forest as a nesting habitat, and Mad Island's patches of low coastal scrub woodland are inadequate for that purpose.

Male Painted Buntings show a splashy patchwork of bright colors, whereas the females are plain pale green. The species winters in Mexico, Central America, south Florida, and the Caribbean, hiding in scrub woodland and seen most frequently when visiting a backyard feeder. For birders, the male Painted Bunting, which sings a musical song reminiscent of that of the more widespread Indigo Bunting, is one of the most sought-after species of the Deep South. The Dickcissel, with its black bib, yellow breast, and chestnut wings, is one of the omnipresent songbirds of the prairie, giving its staccato six-note song from atop shrubs all day long. For birders visiting southeastern Texas for the first time, the sight of these four species out in the open prairie habitat is entrancing.

BIRD-BANDING

Called ringing in Europe, bird-banding has been a means to study bird movements since 1803, when John James Audubon tied silver wire around the legs of nestling Eastern Phoebes in Pennsylvania and found two of them back on their nesting site the following spring. Today, 6,500 bird-banders are active in North America. All operate under the auspices of the Bird-Banding Lab at the Patuxent Wildlife Research Center, in Laurel, Maryland. The lab, founded in 1920, is a collaboration between the U.S. Geological Survey and the Canadian Wildlife Service, and it provides the bird-banders with free numbered aluminum bands that they affix to the leg of each captured bird for permanent identification. Each year the lab provides bird-banders with fresh supplies of bands, and in return it receives detailed records of the birds banded over the previous year. The lab keeps computer files on the more than one million birds banded each year and the seventy million birds that have been banded over the past nine decades. Of greatest interest are the "recoveries"—banded birds that have been retrapped at a subsequent time or place. It is recovery data that yields information on the movements of birds across the continent.

Traditionally, banding studies focus on productive "migrant traps"—areas where migrating songbirds concentrate because of optimal

Dickcissel

geography and habitat. For instance, from 1958 to 1969, Chandler S. Robbins, James Baird, and Aaron M. Bagg situated their Operation Recovery program at several coastal locales in the East to track songbird migration, and these initiatives led to the establishment of permanent bird observatories, such as Manomet, in coastal Massachusetts, and Point Blue, near Point Reyes, California. Several universities carry out seasonal bird-banding operations as well, but most birds are banded by hobbyist banders in neighborhood woodlots and regional parklands. Collectively, these groups and individuals have generated invaluable information on species longevity and how large-scale weather patterns influence the seasonal movement of songbirds, both of which help scientists to understand the evolution of successful migratory strategies.

Mad Island's isolated patches of coastal woodland are ideal for spring bird-banding because they draw in arriving migrants, which find the vast expanses of prairie grassland and marshland unsuitable as stopover habitat.

Over dinner on the night of my arrival, I meet the Smithsonian project team: Emily Cohen, a postdoctoral researcher; Tim Guida, project technical officer; and four assistants. These dedicated bird researchers use a spacious, modern hunting lodge donated, along with the reserve land itself, by the Runnells family in 1989. The compound includes a house with a large, elevated porch, a barn, and outbuildings. Floor-to-ceiling picture windows look out across Mad Island Lake, a birdy coastal estuary, and surrounding coastal saltmarshes. Vistas of prairie and marshland spread out before us.

Mad Island is the southernmost banding project for Neotropical migrants approaching the United States. By the time of my arrival, the Mad Island team had documented a couple dozen species of Neotropical songbird migrants—including Indigo Bunting, Blue Grosbeak, and Hooded, Kentucky, and Swainson's Warblers—dropping into its little patches of coastal scrub woodland. Amid the scrub, the researchers had strung mist nets: forty-foot-long black nylon nets spread tautly between two poles. Mist nets act like large spider webs, harmlessly entangling unsuspecting birds who fly into them without ever seeing them. Thus far, only a few birds were arriving daily, and the team was still waiting for the first big wave of migrants.

The little crew was intrepid, working in a habitat with strong winds, relentless sun, swarming mosquitoes, and venomous snakes. Their long days started well before dawn, which sometimes brought the excitement of trembling nets full of new arrivals and other days only the frustration of nets filled with dried leaves blown off the sheltering trees. Each hour the team checked the nets, mounted along narrow pathways cut through the scrub. Every netted bird was carefully removed, placed in a cloth holding bag, and brought to the project tent for banding. All birds were weighed, measured, and studied for fat deposition, age, and sex. From some birds, the team also collected a small blood sample to check for blood parasites and plasma metabolites that could reveal physiological condition. The team also searched the birds for ticks, which they preserved for study by taxonomic and epidemiological experts. After this meticulous treatment, the team released the birds unharmed to continue their northward journey.

This banding camp operates for nine weeks each spring, and by season's end the team knows all the resident birds of Mad Island as well as the passage migrants that have migrated through. The team told me that the main benefit of working on the banding project is the unlimited access that participants have to the Mad Island coastal prairie environment, with its rich diversity of wildflowers, snakes, frogs, butterflies, mammals, and birds.

During my second morning at the banding camp, a low cloudbank hangs on the coast. A Texas Spotted Whiptail, a nine-inch lizard with green dorsal stripes and small, pale spots on its flanks, searches for a patch of warm sand on the sandy access track to the station. These lands have many reptiles and amphibians as well as birds. This one speeds off in a colorful flash upon my approach.

The first bird netted this morning was a Nashville Warbler. Tim Guida carefully banded and measured it, a process that took about five minutes. Done, he extended his arm with his hand lightly enclosing the small yellow, olive, and gray bird, then slowly opened his fingers. The warbler lay on its back for a couple of seconds before sensing its freedom. It quickly flicked its wings and darted off to a low bush a few paces from the banding tent. It preened for a few moments, collected itself, and flew strongly over the canopy of the scrub and out of sight, its aluminum band glinting in the sunlight. The Nashville Warbler is a boreal forest breeder on its way not to Tennessee (despite its name) but up the Mississippi Valley to the North Woods. This bird, and millions like it, would be preceding me northward in the weeks to come. I hoped to encounter more Nashville Warblers singing on their territory in northern Ontario in mid-June.

As the sun rose toward its zenith, the glare off the sandy expanse of the clearing grew intense; I squinted out from the banding tent over the narrow shipping canal to the expanse of Matagorda Bay, where dredge boats were harvesting oysters and flocks of gulls hovered overhead in hope of snatching something edible. The southwest wind started to build until, just before noon, Guida decided to halt netting. An inevitable challenge of working on the coast is the onshore wind that can blow steadily off the water, presenting problems for bird netting even on otherwise pleasant days. Windblown leaves tangle in the mist nets, which themselves lurch around and snag in vegetation. Because they are somewhat fragile, the nets can tear, making them more visible to birds and thus less effective at catching them. Guida instructed the team to close the nets, which they secured with colorful strings.

Nets closed, the team and I went out to explore the preserve, walking on sandy roads through the property and visiting Skeeter and Pintail ponds. We saw many resident marsh birds but were surprised by what we found far from any woods: two male Orchard Orioles and a Wood Thrush, holed up in a single isolated shrub. This is the type of place an exhausted songbird migrant ends up when it descends into an inappropriate habitat after a long trans-Gulf flight. At least these birds had made it to terra firma, done with their crossing of the open sea.

A pair of White-tailed Hawks put on an aerial show for us over the low grassland—the climax of the afternoon. They have the bulk of a Red-tailed Hawk but unmarked all-white underparts, a gray dorsal surface, red-brown shoulder patches, and a broad white tail marked with a neat black subterminal band. The species flies with its wings tilted upward, distinguishing itself from a distance by this unusual silhouette. This big nonmigratory tropical raptor reaches the United States only in the coastal prairie of southeastern Texas, but its global range extends all the way to southern Argentina. Here in southeastern Texas, it is a local coastal prairie endemic—one of those species confined to a tiny corner of the country. For that reason, few American birders ever see it.

The next morning, Cohen and the preserve manager, Steve Goertz, continue my behind-the-scenes tour of the preserve. The coastal prairie at dawn is very birdy: a pair of Crested Caracaras hang out in a low tree that held their nest last year. Strongly patterned and wary, the pair makes off before we get close enough for a photograph, exhibiting in flight their strange silhouettes, with long necks and steady, rowing wingbeats. Several coveys of Northern Bobwhite quail race across the dirt track in front of Goertz's truck as we slowly rumble over the prairie, and a Loggerhead Shrike perched on a low fence line speeds off with whirring wings that flash black and white in the morning sun.

The relative abundance of Northern Bobwhites and Loggerhead Shrikes here in southeastern Texas was a pleasant surprise because they have disappeared from virtually all their former range in the eastern United States. Why are they common in southeastern Texas? Biologists think it is due to the abundance of invertebrate prey available here for these two open-country birds. Solutions to large-scale conservation problems can arise in environments such as this one, where a juxtaposition of beneficial environmental factors and benevolent human interventions produce a positive result. Yet finding such solutions is a matter of focused trial and error. No one had ever restored a Texas coastal prairie before Goertz and his small team initiated their program, deploying selective grazing and judicious use of fire and herbicide to control tenacious invasive plant species. The restoration of Mad Island as a native coastal prairie will require decades of effort. But by saving this land from development, the Nature Conservancy has already taken a major step for its protection.

The highlight of the third morning at the preserve was a pair of Upland Sandpipers that flew up from a patch of short grass, giving their weirdly musical, trilled, rising and falling flight song. These grassland specialists had recently arrived from their wintering grounds in the pampas of Argentina. Because of declines in grasslands, eastern birders don't get to hear this song much anymore; today the species is mainly found in the prairie country of the northern Great Plains.

Upland Sandpiper

THE NATURE CONSERVANCY'S PRESERVATION
OF PRIVATE LANDS

Since its founding in 1951, the Nature Conservancy (TNC) has focused on preservation of private lands in the United States. To date, TNC has ensured protection for more than twenty million acres of private lands and managed these properties to maintain or enhance their ecological value. Because of its strong national reputation, TNC often receives donations or bequests of private land of significant natural value, and the Runnells family's land was one such property. While donations such as this one are an effective method of achieving conservation if resources are available, long-term costs are incurred in the maintenance and management of private lands. Sometimes TNC simply purchases land outright and then passes the property to a state or federal entity to carry out long-term management.

Aside from the purchase and maintenance of property, TNC has also pioneered conservation easements on private properties that both benefit landowners and ensure perpetual protection of natural values. Easements enable land managers to achieve targeted objectives while keeping land under private ownership, and they can help conserve tracts of forest, protect water quality and scenic vistas, and create wildlife habitat. Typically, TNC purchases easement rights from a land-owner and thus achieves a conservation "win," while the landowner receives a tax benefit. The United States has a large network of local, state, and national protected lands, but by adding a portfolio of private lands, TNC has expanded the reach of conservation across the country. Today more than 40 million acres in the United States are covered by private conservation agreements, compared to 10 million acres under state protection and 170 million acres conserved by the federal government. Private protection is particularly important for preserving lands with very specific conservation values that otherwise aren't pro-tected by the governmental sector—as is true of Mad Island's critically endangered Texas coastal prairie.

The 2015 netting season at Mad Island lasted from mid-March to mid-May. The team banded 1,972 individual birds of 81 species, including local residents and Neotropical migrants. The season's highlight was a big arrival of Dickcissels, which for several days perched atop virtually every bush, plus rare species including ten Golden-winged Warblers and an out-of-range Western Tanager and Yellow-Green Vireo. Overall, the team observed 230 species of birds during the spring. The 2015 season was memorable to the field team because it was particularly wet, generating abundant wildflowers and many Cottonmouth snakes but no huge flights of Neotropical migrants. No matter the spring weather, though, the team collected another year's worth of spring migration data. As results from more years become available, bird-banding projects such as the one at Mad Island will help us to refine our understanding of northbound songbird migration under various weather regimes and wind patterns.

My stay at Mad Island gave me a small taste of the songbird migration to come. I learned about the coastal prairie and saw first-hand private lands conservation in action. I was now headed up the coast of Texas to another birding hotspot, at a time when birds would start flooding into the coast.

THREE

The Coastal Oak Woods of Texas and Louisiana

MIGRANT MAGNETS

Early April 2015

All of a sudden a warbler starts and stops. All of a sudden it flashes from branch to branch, peers under leaves, snaps up caterpillars, darts on again.

—EDWIN WAY TEALE, *North with the Spring*

In the hour before dawn, a Great Horned Owl hoots its low, cadenced five-note series close to the Mad Island lodge. As sunrise approaches, fog cloaks the prairie, and dew soaks the grass and my tent. The winds are light this morning, so work will go forward at the banding station. There has been no big arrival of Neotropical migrants, and by midmorning, I'll say my good-byes and head three hours northeast toward my next destination: High Island, Texas. Perhaps a big pulse of Gulf-crossing migrants will show up there in the next few days.

igh Island is a tiny coastal community that sits atop a salt dome amid a broad expanse of saltmarsh in the easternmost corner of Texas. It is isolated from I-10 and the strip-mall town of Winnie to the north by the shipping channel of the Intracoastal Waterway, which carries barge traffic between industrial facilities in southern Louisiana and coastal southern Texas. This channel also passes in front of the Mad Island banding camp, three hours to the southwest.

Birding and fishing are the primary attractions of High Island and its environs; its small downtown includes only a single motel and a gas station/convenience store, both crowded in spring with birdwatchers who come here to check out arriving migrant songbirds in the woodland reserves, to visit the local waterbird rookery to see displaying egrets and spoonbills, to wander the adjacent Bolivar Peninsula to spot birds of beach and estuary, and to visit adjacent Anahuac National Wildlife Refuge for marsh and open-country birds. For a few weeks each spring, this corner of Texas produces some of the best birding on earth.

Like Mad Island, High Island is not a true island surrounded by water. Instead it's a small wooded rise ringed by marshlands, its uplands beloved by birders for the oak woodlands that lure passing songbird migrants in spring. The songbirds arrive at High Island after their overwater crossing and descend into the community's woodlots to feed, drink, bathe, and regroup. Winding sylvan trails in the small reserves allow birders to approach these normally elusive birds up close.

OPPOSITE: Northern Parula

High Island is diminutive and its woodland reserves are tiny as well, yet they can attract remarkable concentrations of songbirds. The community features five private woodland sanctuaries owned and operated by the Houston Audubon Society and the Texas Ornithological Society, each of which offers critical food and shelter to migrating birds on their way north. They are the focus of an annual springtime pilgrimage by birders from all over North America, who come in hopes of seeing a songbird fall-out.

SONGBIRD FALL-OUTS

Whereas the Mad Island coastal woods attract both trans-Gulf migrants and those traveling north along the eastern coastline of Mexico, most songbirds arriving at High Island have flown over the Gulf. Many of them depart from the Yucatán Peninsula in southeastern Mexico, about six hundred miles south-southeast of High Island. Some have already flown there from Amazonia, Colombia, and various parts of Central America, journeying in jumps that in some places required them to travel over the Caribbean. On the Yucatán, the migrants feed, rest, and wait for benign southerly winds and fair skies—conditions favorable for a flight north over water.

When conditions on a spring evening are promising, these birds rise into the sky and fly north, eventually reaching a cruising elevation of several thousand feet, depending on where they find favorable winds. Some species apparently fly north in small groups, staying together for the long flight. They keep a fixed course and travel all night, not making landfall on the U.S. mainland until late in the afternoon of the following day: a flight of some fifteen to eighteen hours. Those crossing the Gulf include not only Neotropical songbirds but also birds of various other lineages—herons, ducks, shorebirds, cuckoos, and more. Between mid-April and mid-May, rivers of birds pour northward across the Gulf, millions and millions of them heading toward the U.S. coast. They are not attempting to arrive in a specific spot on the U.S. mainland; nor do they need to. Examining a map of the lands surrounding the Gulf shows that a bird departing the Yucatán and flying generally northward will eventually make landfall somewhere, no matter how far off course it strays, because there is Florida to the northeast; Louisiana, Mississippi, and Alabama due north; and Texas to the northwest. This long arc of coastline can generously accommodate the migrant birds no matter which way the winds may carry them.

If the weather remains fair and the winds are following, the birds have an easy flight, and, upon reaching land, most keep flying northward until they reach the extensive bottomland forests of the mainland interior. If, on the other hand, the winds turn or rain or thick fog disturbs the birds' northward travels, they may have difficulty making it to the coast. On those days, the exhausted migrants tend to descend at the first sign of land and head straight for the nearest coastal woodland patches. This is when places such as High Island prove both vital for the birds and exciting for birdwatchers.

The most extreme bouts of contrary weather create what is called a fall-out, in which massive numbers of northbound birds literally fall out of the sky to land on the coast. A fall-out—while not necessarily killing birds—is a taxing event for them. Each time a migrant songbird heads north from the Yucatán, it is taking a chance. It makes its move because of favorable local weather conditions, but it cannot predict what weather may occur en route. A strong cold front could be heading southward out of the Great Plains. Or a band of thunderstorms might blanket the Gulf Coast. Both spell trouble. Birders watching from beaches during bad weather have seen migrants moving heroically just above the wave tops, fighting a headwind to cover the last few hundred yards to solid ground. Some individuals drop into the waves just short of their destination; other birds make it to the beach and then expire from hypothermia and starvation. Fortunate others labor to shore, land in a shrub, and immediately begin refueling on tiny gnats and other prey.

Offshore observers on oil-drilling platforms have reported thousands of small songbirds flying into headwinds but making virtually no progress. Some stopped and rested on the platforms' superstructure, but most continued on without pausing. It seems that the birds are focused single-mindedly on making it to the coast. Luckily, such cata-strophic events have been seen only rarely by platform-based observers. On most days (and nights), the migrants move at high elevations over the rigs and continue northward, making their crossing successfully.

Why do at least sixty-five species of Neotropical migrant songbirds take the trans-Gulf route rather than the more circuitous one around the bend of the Mexican coast (chosen by seventeen of the species)? Despite the hazards, the overwater route must be the more efficient and advan-tageous one. A migrant takes the overwater journey across the Gulf because it gives the bird a leg up in the race north to claim a breeding territory and optimize, over its lifetime, its production of offspring, con-tributing to the gene stocks of future generations. The unsympathetic hand of natural selection drives the evolution of bird behavior, including the innate selection of particular migratory pathways.

Christopher Columbus and other early explorers traversing the Gulf of Mexico noted the passage of land birds far from shore in spring and fall, but no naturalist took such comments as proof of a trans-Gulf migration route for several centuries. In the 1940s, George Lowery, of Louisiana State University, was the first to argue that songbirds actually do cross the Gulf in migration, but prominent naysayers ridiculed his hypothesis. Lowery worked in the Yucatán at night, using a telescope to spot the dark shapes of birds crossing the face of the full moon on their way northward. Large numbers of songbirds, he discovered, were making the trans-Gulf spring crossing. Those few ornithologists and birders along the Gulf Coast of Louisiana who had witnessed spring fall-outs in Louisiana were convinced by Lowery's reports, but others doubted that such small birds could fly so long without rest and food.

Sidney Gauthreaux, one of Lowery's graduate students at LSU, used pioneering nighttime radar studies in the 1960s to confirm that, yes, indeed, birds migrate north across the Gulf in numbers, often high in the sky. Most recently, with the aid of ever-better weather radar technology, Gauthreaux has provided strong baseline evidence that the number of songbird migrants crossing the Gulf has declined substantially during the forty years he has been conducting this work.

As I traveled northward, I would learn from various experts about the complex causes of the migrants' decline, which are not related to the rigors of their Gulf crossing.

On my first morning at High Island, I bike to the Houston Audubon Society's Boy Scout Woods reserve to purchase an entry badge. Near the registration desk, a solitary male Hooded Warbler, in bold yellow and black, gaily bathes in a shallow bird bath beneath an artificial water drip. Facing the bath is a small grandstand occupied by about a dozen birders who tote binoculars, field guides, and digital SLR cameras fitted with telephoto lenses. Most High Island woodlands feature such drips—and nearby observer benches—to attract migrant songbirds to drink and bathe, and, in turn, to draw groups of birders.

Neotropical migrants were scarce at Boy Scout Woods that first morning, though plenty of local resident birds were in full voice. Still early in the migration season, the weather had not forced many northbound migrants down into these coastal woods; indeed, the little woodland patches of High Island are often migrant-free early in the season. On the other hand, during peak high season, around April 20, a birder does not need a fall-out to enjoy a day of birding far superior to any back at home in Indiana or Maryland; *some* percentage of the migrating birds always drops into the coastal reserves instead of continuing inland.

For those used to Mid-Atlantic birding, the remarkable thing about the arrival of songbirds along the Gulf Coast is that it takes place in the afternoon, not in the predawn hours. A woodland silent at 9 a.m. or 1 p.m. might start swarming with birds at 4 p.m., and birders can experience first-hand the phantomlike arrival of the migrants over the water in full daylight. Today's smart phones and sophisticated weather-tracking technology give birders tools to communicate among themselves and predict where birds will show up, but pinpointing arrivals of big numbers of migrants on the coast remains the realm of guesswork. Birders must venture out to see for themselves what has come in from the Gulf.

SONGBIRD WOODS

A few days later, I visited High Island's Smith Oaks Sanctuary, with songbirds aplenty despite the absence of fall-out. The afternoon show began in the parking lot, where a small mulberry tree in full fruit was the target of two long-lensed photographers capturing shots of an array of migrant songbirds. Several Rose-breasted Grosbeaks ate with gusto, finding the purple berries irresistible; the adult males, which had wintered in a Central American forest, sported a rosy bib against a white breast and belly, harlequin black-and-white upper parts, and a big, triangular pinkish-white bill. On its breeding grounds in the hardwood forests of the Northeast, this grosbeak is a shy canopy dweller, heard but rarely seen, but here one could stand on the lot's grassy verge within fifteen feet of the birds as they foraged in the small mulberry at eye level. Also gorging in the mulberry were a male Orchard Oriole (black and chestnut), a male Baltimore Oriole (black, white, and fiery orange), several male Summer Tanagers (orange-red, with a yellow bill), a male Scarlet Tanager (deep red, with black wings and tail), and several Cedar Waxwings (tan and black-masked, with red and yellow highlighting on wings and tail). Here were five of the most colorful songbirds in North America—all in a twenty-foot mulberry tree by a crowded parking lot.

I was joined by Texas Parks and Wildlife Department ornithologist Cliff Shackelford and his wife, Julie, director of Texas programs for the Conservation Fund (a nongovernmental organization similar to the Nature Conservancy). Both have devoted their lives to the protection of natural habitat for Texas's birds and other wildlife; effective nature conservation, here as elsewhere, is driven by productive, can-do people such as the Shackelfords, who bird every chance they get when they're not working. Julie told me that the Conservation Fund had just helped purchase the Powderhorn Ranch—five thousand acres of remnant coastal prairie adjacent to Mad Island—thus substantially expanding the protected coastal prairie facing Matagorda Bay.

Rose-breasted Grosbeak

In the oak woods, crisscrossed with winding and shadowy trails, we observed fifteen species of passage migrant wood warblers—Cliff's favorite birds and my quest group—over a ninety-minute period. A Black-and-white Warbler crept up a sloping branch, acting like a nuthatch. Below it, an Ovenbird searched fallen leaves on the ground for insect prey. High in the leafy branches of a big old Live Oak crept several warblers—a Northern Parula (diminutive but colorful), a Black-throated Green Warbler (with a black throat patch and yellow face), a Blackburnian Warbler (with a deep-orange throat that seemed to glow), and a male Blackpoll Warbler (patterned a bit like a chickadee, but with yellow legs). Seeing the warblers as they foraged high in the leaves was no easy task. Clumps of birders stood about, helping one another pinpoint the different species and speaking in quiet tones as they compared notes and asked about the whereabouts of certain target species ("Anybody seen a Goldenwing?").

The silence of the passage migrants made them difficult to locate. Because wood warblers are, of course, famed for their singing ability, it was a major surprise to learn that here in Texas, northbound passage migrant warblers only rarely, if ever, sing. Instead, in the High Island oak woods, local residents—Gray Catbirds and Carolina Wrens—gave voice. Another surprise for a first-time birder in the High Island wood-lots was that the warblers, arriving in the afternoon, mostly disappeared northward after a stopover of a just few hours. By contrast, in the Mid-Atlantic environs, where I had done most of my birding, migrants tended to stay two or three days in a patch of woods before undertaking their next flight northward. On High Island, birds dropped in to feed and bathe but flew off northward shortly after darkness fell, in a rush to reach more productive bottomland forest in the interior.

The warblers at Smith Oaks this afternoon were all passage migrants—fun to see and good practice for the days to come, but they didn't count toward my warbler quest. So far, I had racked up only one warbler species on its breeding habitat: the Common Yellowthroat, which I had observed at Mad Island on its nesting territory. One down. Thirty-six to go.

CONSERVATION AND RESEARCH EFFORTS

Smith Oaks and Boy Scout Woods form the epicenter of birding on High Island. Owned and managed by the Houston Audubon Society, these two adjacent preserves exemplify the best in volunteerism and private philanthropy on behalf of nature. Based in Houston and citizen-led, the Houston Audubon Society has established seventeen bird sanctuaries in the greater Houston–Galveston area since its founding in 1969, including several woodland sanctuaries on High Island. Houston Audubon's regional network encompasses more than three thousand acres protected for birds, including Bolivar Flats, southwest of High Island and famed for its beach birds.

Houston Audubon's High Island sanctuaries are operated entirely by volunteers, most of whom drive daily from their homes in Houston to help out. They manage a visitor center, provide information and guidance to the approximately ten thousand nature-loving visitors who come each year, and maintain the properties and their trails, buildings, blinds, drips, and observation platforms. In 2015, a hundred volunteers donated forty-five hundred hours of their time. The sanctuaries were initiated with a purchase of just four acres by this Audubon Society in 1984, and subsequent purchases and donations by citizens and Amoco Production Company have expanded the properties to their current size—a big deal for birdwatching, bird education, and bird conservation, and all of it powered by local volunteers.

Hundreds of citizens' organizations throughout America, including this one, promote conservation, education, and nature study, making life better for migratory birds and more interesting for local people. Yet they are just one type among numerous institutions working on behalf of migratory birds and their habitats. Citizens' groups such as Houston Audubon, state agencies such as the Texas Parks and Wildlife Department, nongovernmental conservation organizations such as the Nature Conservancy, corporations such as Amoco Production Company, and statutory research entities such as the Smithsonian Migratory Bird Center form part of the picture, but in my journey I also saw the contributions of federal agencies and universities as well as those of national wildlife refuges and state parks, along with intriguing partnerships between citizen groups and corporations that are yielding substantial conservation successes.

Late one day, I drive to the landward side of High Island, returning to Smith Oaks Sanctuary. I stop in a dirt parking lot surrounded by low woods and packed with late-model cars sporting window stickers pledging allegiance to diverse birding and nature organizations. With camera and long lens in hand, I follow trail signs to the sanctuary's waterbird rookery, situated on a narrow island in Clay Bottom Pond. Before I take many steps, I hear the cacophony of long-legged waders in full breeding mode.

Amoco Production Company donated Clay Bottom Pond and its enclosed island to Houston Audubon in 1994. Both are artificial, the products of industrial activity as well as water management for High Island. When Audubon took ownership, there was no waterbird rookery here, but after a year of habitat restoration and protection, herons showed up. By 1995, fifty pairs of herons nested on the island. By 1998, thirteen thousand birds of various species used the island as a roost. If anything exemplifies the impact of smart conservation planning and action, it is the creation of this safe space for large waterbirds. Set aside appropriate habitat, protect it well, and the wildlife will come.

Houston Audubon constructed a series of observation platforms on the far side of the narrow water passage opposite the nesting island; patrolled by alligators, the passage keeps out pesky predators such as raccoons and opossums, which might consume the eggs of breeding waterbirds. I spent the next hour, with the sun dropping toward the horizon behind me, gazing in amazement and shooting photographs of the crazy commotion on the island. Scores of Great Egrets, Snowy Egrets, Roseate Spoonbills, and Neotropic Cormorants were scattered over the island; big, brightly colored birds were everywhere, some posing, some carrying sticks for nests, some marching about. Pairs courted, and some birds challenged each other in a swirl of nest construction, territorial aggression, display, and sex.

The Great Egrets in particular were stupendous. One male, a lanky and graceful large white bird, raised the gauzy plumes (or aigrettes) on its back and flanks into a wispy tutu and, as its luminous feathers waved in the breeze, snaked its neck up over its back, pointing its beak skyward and then bringing it forward and downward in a long bow. Carried only during the breeding season, a male's aigrettes are the height of refined and filmy beauty. The delicate, snowy plumes, glowing in the sunlight, looked much like those of a bird of paradise; not surprisingly, fashionable women lusted after these feathers for ornamenting their oversized hats back in the 1890s.

Second in abundance to the Great Egrets were the Roseate Spoonbills. Today they were busy fighting for mates, which they did by dueling with their absurd-looking spatulate bills and waving their pink wings. As they battled, they showed off bright patches of color: an all-white neck; all-pink wings; red flashes on the shoulder, rump, and undertail; and rich ochre at the base of the wing and on the tail. The pale, creamy-green skin of the spoonbill's bald crown, its orange eyes, and its dark-pink legs all added to the extreme effect. The birds, wildly colored and strange-looking, were gorgeous but also slightly grotesque, more striking than beautiful. Roseate Spoonbills are graceful on the wing, but when a pair battles for a nest site or a female, they're like two clowns going at it in a circus.

It is shocking to think that in the year 1900, the U.S. populations of both the egret and the spoonbill were virtually exterminated by the commercial plume trade. Frank Chapman wrote in 1904 of the situation in Florida: "I have heard a 'plume hunter' boast of killing three hundred herons in a 'rookery' in one afternoon. Another proudly stated that he and his companions had killed one hundred thirty thousand birds—herons, egrets, and terns—during one winter."

Today, because of the good work of organizations such as Houston Audubon and the largesse of corporations such as Amoco Production Company, birders may take the spectacle of waterbird abundance for granted. But if we go back 110 years, things would look quite different. The National Audubon Society was founded in 1905 in large part to address a national crisis: long-legged wading birds (herons, egrets, spoonbills, ibis, and others) were overhunted, mainly to adorn those oversized hats for the fashion-conscious. At the same time, broad-scale and unregulated market hunting of all manner of "game birds" was leading to the disappearance of populations of ducks, geese, swans, shorebirds, and even some songbirds. During this period, the Carolina Parakeet and Passenger Pigeon faded into extinction, mainly because of unrestrained year-round shooting. At the end of the nineteenth century, everything was fair game, not only birds: the Bison was nearly extinct, and hunters all but exterminated White-tailed Deer and Wild Turkey from the remnant forests of the East. Market hunting and the extensive deforestation that took place during and after the Civil War were a one-two punch that reduced wildlife to a shadow of what it had been when the Pilgrims arrived on these shores.

At the eleventh hour, Theodore Roosevelt, John Muir, and George Bird Grinnell spoke out on behalf of protection for threatened species, as well as land conservation, and they founded the Boone & Crockett Club, Sierra Club, and Audubon Society. The last of these focused initially on protecting birds of all sorts through the creation of sanctuaries as well as passage of local and federal laws protecting birds and making the commercial sale of migratory birds illegal. Market hunting and plume collecting were fully banned by 1920, but decades passed before the deeply depressed populations of many species rebounded to the levels we appreciate today. It is remarkable to report that many bird populations—especially waterbirds and raptors—are in better shape today than at any time in the past century. Thanks are due to effective legislation, the creation of sanctuaries, and the natural regeneration of forests on unproductive lands that were abandoned by family subsistence farming in the early decades of the twentieth century.

BIRDS OF ESTUARY AND BEACH

The next morning I turn my energies toward meeting friendly birders, introducing myself to the volunteers who operate the Houston Audubon Society visitor center at Boy Scout Woods. Running into three young leaders of Tropical Birding, a nature-tour group, I learn they lead free bird walks each weekday, and I join them on a trip to the coast. Fifteen of us, plus the guides, caravan to Rollover Pass on the Bolivar Peninsula, nine miles southwest of High Island. Here a small bridge spans a narrow outlet draining Rollover Bay into the Gulf of Mexico. Just north of the bridge is a sandy access road to the bay and, depending on the tide, an abundance of sand flats that attract myriad waterbirds during the late winter and early spring. The guides quickly set spotting scopes on tripods and begin pointing out bird species to the birders—some novices, others experienced, but virtually all of us fifty-five or older. For many of us, recently arrived from the wintry North, this is birding nirvana.

Although the woods of High Island were quiet, the flats swarmed with waterbirds. Eight species of terns rested nearby in flocks or foraged for tiny fish in the shallow bay: Least Tern, Black Tern, Common

Tern, Forster's Tern, Gull-billed Tern, Sandwich Tern, Royal Tern, and Caspian Tern. I had seen all of these species at one time or another, but I had never seen the whole lot all at once and all together. In addition, an impressive group of sandpipers and waders assembled on the flats: American Oystercatcher, American Avocet, Black-necked Stilt, Willet, Long-billed Curlew, Marbled Godwit, and a half-dozen smaller species. Five plover species hunted the flats with their distinct stop-and-start gait: Black-bellied, Wilson's, Semi-palmated, Piping, and Snowy. A hundred Roseate Spoonbills and several Reddish Egrets added to a scene that was much like watching the color plates of a North American birding field guide come to life before us.

Not only were there birds in dense profusion, but many perched right at the edge of the water or in the shallows at close range. Long lenses were drawn and close-ups of all sorts of accommodating waterbirds were rapidly uploaded to storage media of expensive digital SLR cameras. The birders of our little group—many of whom were adding new species to their life lists—quietly uttered a sort of ecstatic gibberish as they moved from one handsome bird to another.

Most remarkably, Rollover Pass is not a reserve of any kind. It is just a particularly fertile estuary where many kinds of habitats converge: seashore, tidal pass, sandflat, and marshland. That said, the profusion of birds here is a result of the nearby array of protected areas scattered in almost every direction. This section of coastal Texas is rich in conservation green spaces, and the birds that breed and roost at night in these areas visit Rollover Pass when the tides produce good meals.

Rollover Pass's parking lot was filled to capacity, and people of all ages were out and about. They fished, waded in the shallows, dug clams, and lazed by the water—humans and birds and sunshine mixing in an happy outdoor tableau. Of course, the birding group stood out from the others: most of us were gray-haired and swathed in baggy clothing made of drab-olive miracle fiber sold at great expense by online outdoor outfitters. We wore wide-brimmed floppy hats, some with French Foreign Legion–style sun protectors draping over the neck and onto the back. Compared to the clammers nearby—barefoot, in shorts, mostly shirtless—we birders seemed almost a distinct species.

After about an hour of high-octane birding, our leaders pulled the plug. They were hungry and decided to head to a famous taco truck down the road at Crystal Beach. I followed. The cilantro-flavored beef tacos were as stunning as the birds at Rollover Pass, and we washed them down with sweet, mandarin-flavored Mexican soda.

Afterward I drove down the peninsula to Bolivar Flats for beach birding with a conservationist twist. I was scheduled to meet Kacy Ray of the American Bird Conservancy (ABC) and learn about her team's work conserving beach-nesting birds, which they carry out on the Gulf Coast from Florida to Texas. Ray partners with local organizations to deploy paid field staff and volunteers who address local threats to the Wilson's Plover, Snowy Plover, Black Skimmer, and Least Tern, comely waterbirds that build nests in the sand just above the high tideline and thus are vulnerable to beachgoers, unleashed dogs, motorized vehicles, and predators such as raccoons, coyotes, and gulls. The work must be carried out every spring and summer, year in and year out. Ray is a conservation warrior on the front lines, and her fighting spirit is admirable.

Today Ray planned to visit Kristen Vale and Stephanie Bilodeau, ABC–Houston Audubon shorebird technicians who were banding Wilson's Plovers on Bolivar Flats in order to monitor and conserve this declining species. For each such threatened beach-nesting species, Ray's teams locate and monitor nesting sites, fence them off to keep the public and beach drivers from destroying the nests, address predator issues, and conduct public outreach to educate beachgoers on how to avoid harming the nests and young during breeding season.

I watched Vale and Bilodeau trap and color-band nesting pairs of Wilson's Plovers on a stretch of sand above the tideline, using a simple box trap adjacent to the nest that was tripped by a string pulled by Vale when a bird passed underneath (talk about low-tech). By banding these birds with unique color combinations, they can then identify individuals, better estimate the total number of birds, breeding pairs, and territories, and assess long-term survival and interregional movements.

ABC, founded in 1994 and built upon productive partnerships such as the one Ray is implementing, today spearheads an array of innovative programs that conserve native bird populations in the Americas, addressing threats posed by habitat destruction, pesticides, feral cats, wind turbines, urban lighting, window strikes, and more. Moreover, ABC has driven the nationwide Partners in Flight collaboration—a network of scores of government and nongovernment institutions working for conservation of migratory songbirds. Later in the spring, I would visit an ABC field project in Minnesota that is creating new breeding habitat for the Golden-winged Warbler, another Neotropical migrant under threat.

ANAHUAC NATIONAL WILDLIFE REFUGE

The next morning, I find that Boy Scout Woods is again migrant free. I decide to head to Anahuac National Wildlife Refuge, just northwest of High Island, where I'm joined by Jared Keyes, an ace birder with a sharp ear whom I know from springtime birding in another famous migrant trap: New York's Central Park. We drive slowly along the refuge's wildlife loop south of the visitor center. It's noisy with common species: Boat-tailed and Great-tailed Grackles, Common Gallinules, Neotropic Cormorants, Red-winged Blackbirds, Marsh Wrens, and Savannah Sparrows. The loop passes through freshwater wetlands filled with various marsh grasses and reeds. In an East Texas April, this sort of habitat bursts with birdlife—birds are everywhere, perching prominently and vocalizing. But we're hunting hard-to-find rare species.

"Least Bittern!" Keyes called out, and I jerked the car to a stop at the edge of a waterway. We happily glassed the bird, the smallest and most reclusive of the herons; honey-brown, with a blackish cap, the tiny species is a treat to encounter because of its rarity. This individual's dark back indicated it was an adult male. We watched the bittern clamber gingerly among the reeds, hunting small aquatic vertebrates. Perhaps it had just arrived here from a winter sojourn in Mexico. I had not seen this species in four decades, so I was most pleased with my tripmate, who was proving to be an excellent spotter.

I drove less than a mile more before Keyes yelled again. "Stop the car!" He had heard the call of a Black Rail. I veered to the side of the gravel road, and we both hopped out. A trilled *kih-kee-kerrd* came from thick marsh grass near a fence line about ten paces from the car. We crept close. The Black Rail is one of those super-elusive species that makes it onto birders' most-wanted short lists; nonmigratory, it is restricted to coastal salt marshes and rarely leaves the cover of thick marsh grass. Neither Keyes nor I had ever seen one.

We stood on either side of the tussock of grass that hid the tiny marsh dweller and waited for it to show itself. It wasn't interested. Our patience depleted after twenty minutes, we headed back to the car. We had communed with the vocalizing little creature from a distance of a couple of yards, and that was OK with us. It's good to leave some birding ambitions to a future date, and hearing this small recluse at close range was almost as good as seeing it.

Our final stop of the morning was Jackson Prairie Woodlot, a tiny strip of trees planted on a sliver of raised ground in the middle of this vast, treeless marshland. We could walk its perimeter in about ten minutes. A few migrant songbirds sheltered here in the late morning—Blue Grosbeak, Indigo Bunting, Black-throated Green Warbler, Blue-headed Vireo, and several Summer Tanagers. Seeing the place's potential, I decided to return on an afternoon more favorable to an arrival of songbirds from across the Gulf.

I returned alone around 3:30 p.m. the next day, keeping in mind how unpredictable the arrival of migrants can be; often it is best to simply go out into the field armed with no more than hope and a pair of binoculars. The weather forecast indicated conditions—fair skies, light southerly winds—favorable to an arrival of migrants from the Yucatán. I worked the perimeter of the woodlot and found small flocks of Indigo Buntings and Blue Grosbeaks flitting around the outer edges of the woods, and warblers and vireos foraging in the shady interior.

Then I looked southward from the southern tip of the woodlot and saw, through my binoculars, groups of buntings and tanagers moving north up the road from the coast. They passed over me and dove into the trees of the woodlot—the only trees within a mile or more. More and more birds appeared from the south, crossing the broad stretch of marshland sometimes twenty or thirty at a time: Summer Tanagers, more Blue Grosbeaks and Indigo Buntings, Rose-breasted Grosbeaks, and Dickcissels. Many of these migrants had just completed their trans-Gulf flight and were making their first landfall. Many of the wood warblers fell into the woodlot from great heights, so high I missed the moment of their arrival. But they were creating their own little fall-out—just what I had come to Texas to witness.

After an hour, the woodlot was vibrating with birds: thirteen species of warblers, two species of orioles, two tanagers, two grosbeaks, Yellow-billed Cuckoos, two wary thrush species, and many flocks of Indigo Buntings. Birds lurked in every tree. Birds shuffled on the ground in the shadows. Everywhere I looked I saw birds—*good* birds. I circuited the woods five more times and with each pass saw dozens and dozens of migrants. Each circuit added new species to my list. The little woodlot was filled to its brim with migrant birds fresh off the Gulf crossing.

Yet the next morning, when I again returned to the woodlot, I found it empty. Like a vessel, the woods had filled with birds the preceding afternoon and emptied over the evening hours as the recent arrivals, rested and fed, moved northward into the interior, bound for more capacious and lush woodlands, perhaps along the Trinity River bottomlands north of here. Thinking back to the preceding afternoon's fall-out, I was elated by the immediacy and excitement of those ninety minutes in Jackson Prairie Woodlot; I had experienced a fall-out first-hand and had witnessed the songbirds coming in off the Gulf.

NUNEZ WOODS AND THE HURRICANE LANDS

From High Island, I drive eastward to the central coast of Louisiana to visit another bird-banding operation, this one at Grand Chenier. In the

brooding weather, the journey is one of overwhelming gloom as I pass amid
the grim artifacts of the petroleum industry and the commercial fishery
along the roadside, set in vast expanses of bayou and marshland with
abundant evidence of hurricane destruction. As I drive, I'm thrashed by
a nasty storm that follows me along the coast—a common feature of the
Deep South in springtime. When I reach Port Arthur, the low black clouds,
flashing lightning, and hundreds of smoke-belching refinery stacks create
an infernal scene. I traverse Sabine Pass and Sabine Lake and cross into
Louisiana by midafternoon. Route 82 eastward takes me to Holly Beach,
where a tiny ferry carries me over Calcasieu Bayou to Cameron. I pass no
cars on the road this afternoon, and I have rarely felt so lonely. To say this is
backcountry is an understatement—Route 82 across the exposed underbelly
of Louisiana is a land that time forgot.

I paused my journey in Creole, where a solitary bright spot awaited
me: the Bayou Fuel Stop general store. Here I sampled two finger-
food Cajun delicacies: boudin blanc sausage and boudin balls, both
composed of ground pork meat and liver, dirty rice, and Cajun spices.
Boudin blanc, the standard Acadian sausage of southern Louisiana, is
stuffed into a pork casing and steamed in a rice cooker; it's called *blanc*
because it lacks the pork blood of boudin noir. Boudin balls are small
spheres of the same ingredients but dipped in batter and deep-fried.
(Those who wish to sample such treats should attend the Boudin
Cook-Off, held every October in Lafayette, Louisiana—try the seafood
boudin, which includes crab and shrimp.)

In the late afternoon, I arrived at Grand Chenier, a small coastal
community about 150 miles west of New Orleans, to visit Rockefeller
State Wildlife Refuge, the base of operations for Frank Moore's migratory bird-banding team in Evariste Nunez Woods and Bird Sanctuary. Moore, a professor at the University of Southern Mississippi in
Hattiesburg, set up the banding project in Nunez Woods in the spring
of 2015 after operating at Johnson's Bayou, to the west, from 1993 to
2014. The program is a collaboration between his university and the

Louisiana Department of Wildlife and Fisheries, and I was one of its first visitors. The privately owned Nunez Woods, managed by the Rockefeller Refuge, is a mile-long strip of hardwoods surrounded by marshland and pasture. As the only substantial patch of forest within miles, it is a target for incoming trans-Gulf migrant songbirds.

Named for its stands of oaks (*chene* is "oak" in French), Grand Chenier is set along an ancient Gulf beach ridge stranded inland by erosion and the historic deltaic processes of the Mississippi River. The ridge's slight elevation encourages diverse upland woody vegetation to take root and thrive here, while it cannot in the adjacent lower marshy areas. As in High Island, here in southern Louisiana the landscape is dominated by marshland and coastal prairie, with only a few small patches of oak woodlands, which act as songbird migrant traps in spring.

Along with Water Oaks, Live Oaks are the dominant tree species at Nunez Woods, providing both habitat for trans-Gulf migrants and important protection from hurricane-driven storm surges and flooding. The best chenier woodlands are filled with Live Oaks, many-branched, pleasing to the eye, and beloved by warblers, which forage in them for insect prey. These trees are perhaps the most important native tree in the culture of the Deep South, not least because of their broad and shady branches, many of which stretch out from the trunk nearly horizontally, creating a wide crown and a network of limbs friendly to climbers. Their thick, small, hardy evergreen leaves form a dome of foliage, mimicking the broad porches seen in small towns across the Deep South, and their shade is a valuable commodity in late spring and summer in the hot zone. Their widespread presence makes them a symbol of the South itself, and their natural "tinsel" of Spanish Moss gives small towns their antebellum look, mesmerizing northern visitors such as myself.

Yet there is another omnipresent symbol in these coastal areas: the harsh scars of hurricanes past. For people inhabiting the Gulf coastline from Galveston east to New Orleans, the names Katrina (2005), Rita (2005), and Ike (2008) bring back terrible memories. There are few locations along this stretch that did not suffer the impact of one or more of these damaging cyclonic storms. Anahuac National Wildlife Refuge and the Bolivar Peninsula took a vicious hit from Ike and are still recovering, which will take decades (not to mention the restoration that will be necessary in the wake of 2017's Hurricane Harvey). The works department can rebuild roads and buildings, but natural habitats recover at their own pace.

Rita, the most intense tropical cyclone ever to cross the Gulf of Mexico, made landfall at Sabine Pass on the Texas-Louisiana border on September 24, 2005. A fifteen-foot storm surge struck the coast of southwestern Louisiana, sweeping away most buildings near the shoreline and flooding a vast swath of low country with saltwater. The Louisiana communities of Cameron, Creole, Grand Chenier, Holly Beach, Johnson Bayou, Little Chenier, and Oak Grove received the brunt of the blow, with 90 percent of their homes, businesses, and infrastructure destroyed. More than a decade later, Cameron Parish communities south of the Intracoastal Waterway are still slowly recovering, their populations greatly diminished from pre-Rita levels.

Ike made landfall near High Island on September 13, 2008, slamming the coasts of the Bolivar Peninsula and Galveston Island and spreading its damage eastward into Louisiana, which was still attempting to recover from the violence inflicted by Rita just three years earlier, as well as Hurricane Gustav, which had struck the states only two weeks before Ike. It is presumably the recurrent destruction wrought by hurricanes that keeps this section of Gulf Coast as low marshland with few woodlands, which are killed by saltwater inundation. Coastal habitat loss is just one more challenge that the trans-Gulf migrant songbirds face each spring.

Refuge wildlife biologist Samantha Collins gave me a tour of Rockefeller, which is mainly marshlands and water impoundments constructed to provide foraging habitat for wintering waterfowl. Encompassing seventy-six thousand acres on the south side of Route 82 and extending to the Gulf, Rockefeller is a popular destination for birdwatching, sport fishing, and recreational crabbing and shrimping. Moreover, the refuge is an important wintering ground for waterfowl. As many as ten thousand Snow Geese winter here, as do thousands of Gadwalls, Green-winged Teal, Northern Shovelers, Ring-necked Ducks, and Northern Pintails. The waterfowl population tops 170,000 at the height of the winter season. No waterfowl hunting is permitted in Rockefeller, but in the surrounding private lands, the birds are fair game.

The refuge manages an alligator breeding program, and staff also oversee the statewide farming of alligators and a managed annual harvest of wild gators. After we visited the covered breeding pens, where dozens of cute young gators bobbed about, we drove down the gravel dikes of the impoundments. The place was alive with springtime birds—I goggled at flocks of White Ibis, waterfowl, egrets, and herons flushed out by the sound of the truck.

Last, Collins showed me Nunez Woods itself—tall, wet, and quite lush and tropical, it reminded me a bit of the New Guinea jungles I'd explored over the course of my career. I had a good feeling about it, even though it now lay quiet before us.

The next morning, I met the bird-banding team—Keegan Tranquillo, Shawn Sullivan, and Lauren Granger—and we caravanned to the Nunez Woods banding camp, on the north side of Route 82. Nunez, a private and gated hunting property, is seasonally open to birders. A mix of field and woods, it's thick with deer stands (as in Texas, hunting and fishing are Louisiana's main pastimes). The forest includes Live Oak, Water Oak, Southern Prickly Ash, Green Ash, Southern Hackberry, American Elm, and Chinese Tallow. Saw Palmetto dominates the understory. These woods, with trees topping seventy-five feet, form a long strip about a quarter-mile wide. In the shaded interior, the bird-banding team cut a network of trails along which they deployed mist nets. I would wander these paths repeatedly over the next few days.

Nunez is a world unto itself. A few local resident birds sang this morning—resident Northern Cardinals and Carolina Wrens as well as territorial White-eyed Vireos and Wood Thrushes, Neotropical migrants that nest here. Virtually none of the northbound passage migrants vocalized. Instead the silent migrants skulked in the shadows, making it difficult to locate them. This raised a biological question in my mind: what mechanism prompts male migrants to sing as they head north to their breeding ground? While birding in the Mid-Atlantic, my experience had been that male migrant songbirds sing while in passage. Yet here in the Deep South, that was not the case. So what *does* prompt migrating males to start singing? It may be a hormonal shift impelled by the position of the sun at the birds' breeding latitude, but future research is needed to understand the onset of male song during the spring passage north.

Surprisingly, a flock of White-throated Sparrows perched in a thicket at the woodland edge. This species breeds throughout the Great North Woods and winters in the Mid-Atlantic; it is a common winter yard bird to many people living in the East. I certainly did not expect to find it at the edge of a swampy subtropical woodland on the Gulf Coast of Louisiana, where a Great Kiskadee called *kik—keweer!* from the forest canopy. I realized that I was, in fact, encountering these sparrows at the extreme southern edge of their *winter* range. The familiar birds were still wintering here while Neotropical migrant songbirds were going north—an interesting kind of two-way traffic. The sparrows would eventually head back north to nest in Ontario and the Adirondacks, but they were in no hurry. They stayed quiet, although I'd later hear them practice their song at points north along the Mississippi, and I looked forward to hearing their haunting five-note whistle once I reached the North Woods.

Tranquillo reported that there had recently been a decent migrant arrival. He showed me netted Kentucky, Hooded, and Swainson's Warbler individuals as well as a colorful male Painted Bunting. I'd arrived at Nunez at the beginning of the high season for songbirds coming across the Gulf. As was the case at High Island, these birds were arriving from their overwater journey in the afternoons—the birds I'd spotted this morning had arrived on an earlier afternoon and were resting and refueling before continuing northward into the heart of the continent. The woodland here (unlike the tiny woodland patches of High Island) was large enough to keep the migrants around for a few days. The netting team captured seventy-five birds that busy day at their banding camp, situated in a screen tent tucked into the southern edge of the woods. Tranquillo processed birds throughout the morning and afternoon, and out in the woods Sullivan and Granger pulled birds from the nets almost continually.

I added a second breeding warbler to my quest list: the Northern Parula. I found the tiny canopy-dweller in a Spanish moss–draped Live Oak in the interior of the woods. This, the smallest wood warbler, is an energetic vocalist that prefers high-canopy twigs for its singing and foraging. Blue-gray, yellow, and white, the species is distinguished by its dark collar, broken white eye-ring, white wing-bars, and the yellowish patch on its back. It is interesting that no wood warblers besides this one breed in the coastal woodlands—the habitat simply must not provide enough breeding-season sustenance to support these voracious insect-eaters. Yet the Northern Parula is one of the most widespread and adaptable of the warblers, breeding in coastal Live Oaks as well as the North Woods boreal forests. I would encounter this little sprite again in various places along my route to Canada.

On my second morning at Grand Chenier, I saw flocks of small, dark birds moving along the woodland edge as I drove down the entrance track to Nunez Woods. There had been a migrant arrival at the end of the preceding day, and now the woods, edges, and grasslands hosted hundreds of Indigo Buntings, White-eyed Vireos, and Summer Tanagers. Several other migrant species were everywhere—Kentucky and Hooded Warblers, Scarlet Tanagers, and Blue Grosbeaks.

Birding in Nunez Woods was fascinating because of its strange ornithological juxtapositions. This morning, two distinct ecosystems intersected: as I stood in the forest, looking at Hooded Warblers on the ground and Summer Tanagers in the canopy vegetation, I could see large waterbirds just above the treetops, winging over our little patch of woods. White Ibises, Tricolored Herons, and Great Egrets were on the move from one large wetland to another. One does not expect to be able to watch furtive wood warblers and colorful herons on the same spot of habitat, but this is the kind of treat birders encounter in southernmost Louisiana.

The Nunez team was excited by the presence of a male Cerulean Warbler foraging prominently in a big Live Oak just around the bend from the banding station. Indeed, blues, as well as reds, were to be the themes of this day. Bunches of all-red male Summer Tanagers foraged in the forest and loafed at the edge of the woods. Flocks of Indigo Buntings, joined by larger Blue Grosbeaks, foraged low in the grass and flushed up into the bushes and trees at the edge of the woods each time I walked by. Nervous flocks of twenty to thirty deep-blue bunting and grosbeak males were joined by the brown females of the two species, and the two-tone flocks started up from the field in a dozen intermittent explosions of deep blue and buff brown. To date on my trip, I had recorded 166 bird species without expending much effort— I'd just put myself at the right places at the right times.

On this trip, as I would discover, flocks of Indigo Buntings followed me up the Mississippi Valley to their northern breeding limit at the Canadian border. This bunting—the male a deep ocean blue and the size of a small sparrow—became my colorful little mascot: I'd encountered small groups of them at Jackson Prairie Woodlot, and there were flocks of dozens here. As a youngster in Baltimore, I had found singing Indigo Buntings on territory in virtually every woodland clearing I explored. I loved their complex and rollicking song—an insistent *sweet-sweet* . . . , followed by four distinct repeated phrases—but, because of their abundance, I'd taken the birds for granted. Now, on this long road trip, I was regaining respect for the bright little songbirds, mainly because of their flocking habit and their omnipresence. Here was a Neotropical migrant, one that wintered in Mexico and the Caribbean, that was truly prospering, and I saluted their success as travelers and widespread North American breeders.

Meanwhile the Nunez Woods banding crew was hard at work. Wearing tall rubber boots, they slogged the woodland trails almost continuously as yet another rainstorm pounded the paths into a deep slurry of soft mud and standing water. The nets were catching lots of migrants and there was no time to shift them to a drier set of trails, but the joy of intimate encounters with the diversity of colorful songbirds dulled the annoyance of the all-encompassing mud. The staff and I were paying a small price for a unique experience.

The data on fat and muscle condition that Tranquillo collected from the netted birds indicated that the migrants had arrived in good shape after their Gulf crossing, and the birds I spotted out and about appeared fine, despite the nasty weather they had encountered. Even the tiny hummingbirds seemed unaffected by the Gulf crossing. Of course, their health shouldn't have been a surprise. Natural selection has operated on these songbirds for thousands of generations, and the results of that process are clear: the trans-Gulf route is their best route north.

I had now completed my southern coastal tour. My count of wood warblers on their breeding grounds remained at two, but I knew that about a dozen species bred in the extensive interior forests of the Mississippi bottomlands, where I was headed next. I had observed along the coast a number of passage migrants on their way to the Great North Woods to breed—Swainson's Thrush, Blackpoll and Magnolia Warblers, and Philadelphia Vireo, among others. The wood warblers would lead me north week by week. Heading north from Nunez Woods, I would next hunt for productive migrant stopover sites in the Deep South's interior. Ahead of me lay the Mississippi's once vast and forbidding ancient bottomland forests, at one time the land of the Cougar, the Red Wolf, and the Ivory-billed Woodpecker.

FOUR

The Low Country of Louisiana and Mississippi

Mid-April 2015

*Each time we crossed a brook among the wooded ridges, on that day
of warblers, we stopped. For there we were sure to find a pocket of
migrants. The trees beside such streams were always filled with the
song of the spring woods, the small and varied music of the warblers.*

—EDWIN WAY TEALE, *North with the Spring*

From the Gulf Coast, my path leads me north into the interior bottomlands of Louisiana and Mississippi, with their mix of hardwood swamp forest, river oxbows, row-crop agriculture, and the tiny old towns of the lower Mississippi. The ecologically rich, junglelike forests of the Mississippi Delta are the places that Gulf-crossing migrant warblers hurry to reach after their brief stopovers in the coastal cheniers. More than twenty species of wood warblers pass through here en route to parts north, and populations of thirteen migrant wood warblers actually stop to breed in these forests. How many breeders will I find, and how many passage migrants will I see? And what environmental conditions will my quest birds face?

Before I departed Grand Chenier, a terrific supercell storm struck, with black skies, sheeting rain, and powerful gusts. Luckily, I had broken down and stowed my tent, but the refuge hostel, where I took shelter, rocked on its tall support timbers. Several nearby trees crashed down. Spring on the Louisiana coast is punctuated by a succession of violent storms that typically pass from west to east, and one never knows in advance if a twister is embedded in a dark and swirling mass of cloud. Although not quite hurricanes, these events are fearsome.

Driving the back roads, I saw that the chenier woodlands east of Grand Chenier had suffered a dieback, rendering the area less welcoming to arriving migrants. The dieback was a product of saltwater inundation from the hurricanes that made landfall here, and decades will pass before these cheniers recover. North of Pecan Island (a town, not an island), I came to cultivated rice fields alive with Fulvous Whistling-Ducks and other waterbirds; I was leaving the vast and lonely expanses of marshland and arriving in inhabited farmlands with crawfish ponds and cattle pastures. Here many farmers cycle their fields from rice to crawfish and back to rice, which can produce bird-friendly wetland habitat. Both depend on seasonal flooding of diked fields, and both provide good foraging opportunities for shorebirds, long-legged wading birds, and waterfowl. Ducks Unlimited (DU) works with farmers in southern

OPPOSITE: Hooded Warbler

Louisiana and southeastern Texas to create waterbird-friendly winter wetlands, an effort partly funded by the U.S. Department of Agriculture through its Natural Resources Conservation Service.

Passing through the decrepit town of Opelousas, I entered the watershed of the Atchafalaya—the river that is trying to capture the lower Mississippi. Where the Atchafalaya passes close by the Mississippi, the Atchafalaya is the lower watercourse. Thus, if the paths of the two meet, the Mississippi will be drawn down into Atchafalaya's lower basin and follow its course to the Gulf. The Army Corps of Engineers has spent many millions to prevent this catastrophe of their own creation (more about the Corps' misdeeds later in this chapter). If the Atchafalaya does capture the Mississippi, Baton Rouge and New Orleans will lose their river and the Mississippi will flow into the Gulf about seventy-five miles west of where it does today. The Atchafalaya, created by the confluence of the Red and Black rivers, meanders in a big, swampy bottomland that I crossed as I headed eastward on Highway 190 toward the town of Lottie. I drove several miles atop a raised causeway passing over the swamp forest of Atchafalaya National Wildlife Refuge.

I needed my GPS to navigate this little rural patch of low country that has been much confused by the periodic shifting of the two big rivers and their various tributaries. In the balmy afternoon, I traveled back roads past the small communities of Blanks, Livonia, Frisco, Parlange, and Mix, finally coming to the prosperous town of New Roads. The country here is pretty: a mix of tall woodland and agricultural fields bounded by neatly planted rows of trees. Cattle Egrets forage in the fields, and it has the feel of Virginia horse country but without any prominent hills. At Pointe Coupee, I crossed the Mississippi on the John James Audubon Bridge, a graceful engineering marvel of concrete and steel completed in 2011. It is the second-longest cable-stayed bridge in the Western Hemisphere. The swirling brown river was in flood, and lots of bottomland was underwater.

On the east side of the river is West Feliciana Parish, home of the historic town of Saint Francisville, just uphill from the ancient community of Bayou Sara, right on the main stem of the big river. I was here to visit Oakley Plantation, where John James Audubon once worked while he was struggling to become America's ornithologist. And, of course, I'd come in search of the various wood warblers that nest in this low country—the same birds that Audubon marveled at nearly two centuries ago.

NORTH AMERICA'S FLYWAYS

The geographic paths followed annually by the low-country wood warblers help to define what is known as the Mississippi Flyway, a route up the middle of the country that is traveled by tens of millions of birds every year. Bird migration north and south across the continent follows natural pathways created by major physical features: the rivers, mountain ranges, and coastlines that link major breeding grounds to wintering grounds. Four flyways have been delineated by the concentrated movement of birds of various kinds in the autumn and spring. In eastern Louisiana, I was currently in the heart of the Mississippi Flyway. To the east lay the Atlantic Flyway, and to the west the Central and Pacific Flyways. Flyways are idealized simplifications of the messy reality of the migratory routes taken by the hundreds of bird species that move long distances between nesting and wintering habitats. Still, the concept is useful, as it highlights the geography and magnitude of the migratory phenomenon.

The **Atlantic Flyway** lies between the crest of the Appalachians and the coastal waters of the Atlantic Ocean, from Labrador to Florida. Birds use this route in many different patterns. Some, such as the Broad-winged Hawk, migrate south along the mountain ridges of the interior. By contrast, the Peregrine Falcon migrates along the Atlantic coastline. Some Hudsonian Godwits travel from Alaska to the coast of New England in autumn and then fly nonstop south across the Atlantic to northern South America. Some shorebirds and waterfowl breed in Alaska and cross Canada before taking the Atlantic route south. The Black-throated Blue Warbler, one of my quest birds, migrates mainly along the Atlantic Flyway and is relatively rare along the Mississippi.

The **Mississippi Flyway** follows the natural pathway offered by the great valley carved by that river's main course. Large numbers of songbirds, waterfowl, and shorebirds use this direct route from breeding grounds in Canada and Alaska to reach wintering areas in Mexico, the Caribbean, and South America. This, of course, is the flyway that features the trans-Gulf crossing. The western verge of this flyway merges imperceptibly into the Central Flyway in the Great Plains to the west.

The **Central Flyway** crosses a mix of plains and mountains as well as arid lands. The route leads from northern Canada south across the intermontane West to Mexico and through Mexico to South America. The Hudsonian Godwit, which in autumn flies east from Alaska and uses the Atlantic Flyway, in spring arrives on the coast of Texas and flies north up the Central Flyway to its Arctic and sub-Arctic breeding habitat. The great flocks of Sandhill Cranes that stage on the Platte River in Nebraska use the Central Flyway in both spring and fall.

The **Pacific Flyway** generally follows the West Coast from Alaska and British Columbia to Baja and southward along the Pacific coast to the Southern Hemisphere. This route is used by Arctic Terns, which breed in Alaska and winter on sub-Antarctic islands south of Chile, a one-way journey in excess of twelve thousand miles. Vast flocks of geese and other waterfowl travel the route to reach wintering grounds in the interior of California and Mexico. Together the four major flyways define the North American bird migration system and exemplify the sheer diversity of the migratory habit in our birdlife.

AUDUBON ALONG THE SOUTHERN MISSISSIPPI

On my first morning in Saint Francisville, I am awakened early by the songs of Summer Tanager, Orchard Oriole, and Great Crested Flycatcher from the oak canopy. I rise and travel the low-country road to Cat Island National Wildlife Refuge, down on the Mississippi. In the bottomland woods I hear the voices of several passage migrants—Nashville, Black-throated Green, and Chestnut-sided Warblers—in counterpoint to the song of commonplace local breeders such as the Carolina Wren, Tufted Titmouse, and Northern Cardinal.

Chestnut-sided Warbler

The verdant bottomland forest of the refuge sheltered singing Kentucky and Prothonotary Warblers, two more of my quest birds, on their breeding territories. The Kentucky has a song reminiscent of the Carolina Wren's, but fuller and less complex. This powerful songster is olive above and bright yellow below, with a drooping black ear streak. This is a true denizen of the deep-forest interior, its breeding range entirely confined to the eastern United States. It winters south to Colombia and Venezuela. Whereas the Kentucky was difficult to spot in the forest, the Prothonotary—the golden swamp warbler—was a flash of orange-burnished yellow with blue-gray wings, often in full view on a prominent perch. Its monotonous *swit swit swit swit swit swit* signaled the presence of a swamp, for this species nests only in trees standing in swamp water—something Cat Island National Wildlife Refuge has plenty of. Because of its strong affinity for swamplands, the Prothonotary is especially common in the Deep South. It winters as far south as northern South America.

After my visit to Cat Island, I drove to downtown Saint Francisville, the most picturesque and historically preserved small town I'd seen in the Deep South. Its main residential street was lined with period bungalows painted white or a pale pastel. All were built more than a century ago, and all had been lovingly preserved: small, cozy, and set under the deep shade of Live Oaks and other old trees. I took breakfast at the Birdman Café, where a resident, seeing my field guide on the table, struck up a friendly conversation about birdwatching. Such conviviality, combined with the fine architectural touches downtown, was beguiling. This was a place where it would be fine to retire—or at least spend the winter.

Afterward I drove back to the campground and bicycled to the Audubon State Historic Site at Oakley Plantation. John James Audubon, who arrived here from New Orleans on June 18, 1821, as an aspiring bird artist, wrote of the site: "The rich magnolias covered with fragrant blossoms, the holly, the beech, the tall yellow poplar, the hilly ground and even the red clay, all excited my admiration." A long entrance drive passed through a mix of grand old-growth pines and towering hardwoods, with Live Oaks and Loblolly Pines prominent among the ancient trees. Greeting me was a morning chorus of breeding birds from the canopy as well as from the thick understory set back from the drive—Kentucky Warbler, Summer Tanager, Great Crested Flycatcher, Wood Thrush, Red-shouldered Hawk, Yellow-billed Cuckoo, and Red-headed and Pileated Woodpeckers were all in voice. I stopped several times to revel in the symphony of spring at this wondrous intersection of ornithology, landscape, history, architecture, and art.

Oakley House, in the early Federal style, is handsomely proportioned and clad in white clapboard. Set in a small clearing at the end of the long, winding driveway, it is distinguished by its two floors of porches and its wooden-slat jalousies, set to block the summer sun. The house is bracketed by several large, spreading Live Oaks, and great Southern Magnolias stand guard near the front porch. Oakley's interior, restored to reflect its appearance at the time when Audubon stayed here, conveys both rural wealth and lived-in practicality. The three-story home contains seventeen rooms, with front and side entrances leading to the landscaped grounds, shaded by oak and Crape Myrtles. It is flanked by formal gardens and several period outbuildings that contribute to the sense of a lost time and place, and a small,

Oakley Plantation, Saint Francisville, Louisiana

understated accompanying museum. The grounds also include a nature trail through the forested reaches of the hundred-acre property. Restoration and maintenance of the estate have been ongoing challenges; Hurricane Katrina blew out many of the upper-story windows and knocked down scores of trees on the grounds in 2005.

Construction began on the house in 1799 when Ruffin Gray, a successful planter from Natchez, Mississippi, moved here onto land he purchased from the Spanish authorities (yes, this still was part of Spain's territory at the time). Gray died before his house was finished, and his widow, Lucy Alston, oversaw its completion. She later married James Pirrie, an immigrant from Scotland, and Eliza, their daughter, was born here in 1805. Eliza Pirrie's educational needs eventually brought Audubon to the household. In the 1820s, Audubon and his family made a living not only through tutoring but also by painting portraits and doing other odd jobs among the wealthy planters of Louisiana and Mississippi, particularly in New Orleans, Saint Francisville, and Natchez. It was in the forests and swamps near the Mississippi that Audubon observed and collected the birds appearing in a number of the color plates of his masterwork, *Birds of America*. Today the house's main rooms feature Audubon prints of birds he painted here in West Feliciana Parish.

In 1821, Audubon took up residence at Oakley as tutor to fifteen-year-old Eliza. His contract required him to spend half of each day tutoring the girl, but the rest of the day he was free to explore the woods and paint the birds he encountered and collected. Audubon's stay at Oakley House did not last terribly long because of a family misunderstanding (it seems Eliza may have become overly enamored of her dashing tutor). Nonetheless, Audubon spent, on and off, more than eight years based out of West Feliciana Parish, and while he was here he painted as many as eighty species for the double-elephant folio *Birds of America*. Saint Francisville, then, was one of the most important places for Audubon as he created his magnum opus. Aside from Oakley, he spent time at Beech Woods Plantation, among other places, and it was during this period that Audubon committed to having his great ornithological opus produced in England, with his painted images reproduced by the renowned engraver Robert Havell.

Bird species illustrated by Audubon in West Feliciana Parish include the Swallow-tailed Kite, Pine Warbler, Pileated Woodpecker, and Red-shouldered Hawk. As highlighted by Mary Durant in *On the Road with John James Audubon*, Audubon and his assistant, George Mason, also illustrated in the book's plates many of the more interesting local plants, including Cross Vine, Jessamine, Toadshade, Red Buckeye, Rose Vervain, and Silver Bells. According to Durant, Audubon himself took little interest in the flora except to liven up his bird compositions, and thus Mason did the lion's share of such work for the master.

After my tour of Oakley House and its grounds, I understood why Audubon loved this landscape. It includes a diversity of natural environments, both upland and bottomland, and the culture and wealth of the plantation families made for a pleasurable lifestyle, something for which Audubon had a taste. Of course, Audubon was also a wanderer, and he traveled much of the length and breadth of the continent, from Key West north to Newfoundland and west to the upper Missouri River. But certainly he undertook the preponderance of his fieldwork and painting in the Mississippi drainage between Louisville and New Orleans.

Just as Audubon stopped over in various homes here in West Feliciana Parish, many species of migrant songbird arrive here in late April and either nest in the area or briefly rest and refuel for the next flight northward. This is important songbird country either way; in every direction from where I stood at Oakley Plantation were tracts of forest teeming with birds in passage, at the height of spring.

STOPOVER ECOLOGY

A number of the warbler species I was following breed in the Great North Woods, north of the headwaters of the Mississippi. Each spring, to get to that distant boreal breeding habitat, these birds need to make a series of flights up the Mississippi Flyway, each flight followed by a stopover of one or more days of rest and refueling in a patch of woods midcontinent. I myself was stopping at the forested sites used as stopovers by the migrating birds.

For a forest-dwelling songbird, each stopover lasts several days. First, while still high in the sky, the bird locates a suitable patch of woods at the end of a night's migration flight. The bird drops into the woods as first light approaches. Once in that patch, it must find a safe roosting place hidden within the trees. It must then find water for drinking and bathing. And, finally, it needs to forage productively by sallying out from its roosting place. Thus the tasks facing the songbird midmigration include predator avoidance, drinking, bathing, feeding, sleeping, and preening. Each stopover may also include competition from resident species as well as other migrants, and there is some indication that stopover patches are sometimes depleted of food resources by the presence of too many feeding birds.

As we have seen, the availability of rich woodland patches along the Gulf Coast for first landfall stopovers is strictly limited and probably on the decrease because of hurricane impacts and coastal development. This makes the interior sites all that more important, as passage migrants need to make up for energetic losses incurred in the trans-Gulf flight as well as the brief coastal stopover. That is why the birds are in such a hurry to get to the interior, where they can feed aggressively and gain the weight needed for subsequent flights north. Much of the lower Mississippi bottomlands have been cleared over the past century for row-crop agriculture, but large swaths of seasonally flooded swamplands remain wooded, with little likelihood of future development. Some of these economically unproductive lands are now formally protected. The big question is whether these extant bottomland forests provide adequate stopover habitat for the tens of millions of songbirds passing northward each year. Is there a habitat bottleneck between the birds' wintering and summering sites? That question has not yet been answered fully by conservation biologists.

LAKE BRUIN AND THE MOUND-BUILDING CULTURES

Departing Saint Francisville, I drive north across the state line into rural southwestern Mississippi and the low, forested Tunica Hills—the southwesternmost vestiges of the Appalachians and only a couple of miles as the crow flies from the main stem of the Mississippi. I visit Clark's Creek Natural Area for morning birding and continue northward toward my next camping spot. Passing country churches nearly every mile, I travel through another of Audubon's stomping grounds—Natchez, Mississippi—before crossing Old Man River back into northeastern Louisiana and into the flat delta lands that hold my next destination, Lake Bruin State Park.

The Tunica Hills of Mississippi are perfect wood warbler breeding habitat: well-watered foothills with tall forest graced with a rich green understory. From bottom to top, this mature forest provides a lush breeding environment for warblers. Nestled in these hills, Clark's Creek Natural Area is almost impossibly difficult to locate but well worth the effort. With deep, shaded glens filled with tall oaks, beeches, hickories, and ashes, it reminded me of the woods around Great Falls, Virginia. Spring butterflies—the Questionmark, the Zebra Swallowtail, the Giant Swallowtail—fluttered in the tree openings, and Summer Tanagers and Yellow-throated Vireos sang. Here I located four more quest birds on their breeding habitat: the Louisiana Waterthrush

Black-and-white Warbler

and Worm-eating, Hooded, and Black-and-white Warblers. My tally now stood at six breeding wood warblers.

The cryptic, sparrow-plumaged Louisiana Waterthrush hunts along the verges of clear-flowing woodland interior streams, wagging its tail as it paces the wet spots in search of insect prey. Its strident and ringing song, which carries far through this environment, starts with clear musical slurs and ends in a jumble of chattering notes. This is the southern species of waterthrush, with a breeding range confined to the eastern United States. It winters from Mexico to Venezuela.

The plain but handsome Worm-eating Warbler is a foraging specialist, searching inside curled dead leaves in the understory for hidden arthropods. Its black-striped head gives it a distinctive look, and its rattled song signals its presence in mature deciduous forest. Breeding in the eastern United States and wintering from the Caribbean islands to Panama, it is another ground nester. Singing sweetly from the higher interior saplings, the Hooded Warbler male sports a yellow face, black cowl, olive upper parts, and bright-yellow breast, and it too breeds only in the eastern United States, wintering through the Caribbean and into Central America. When one finds a patch of forest with both the Hooded and the Worm-eating on territory, this indicates quality forest habitat.

Finally, high in large canopy branches, the creeperlike Black-and-white Warbler sings its lisping song while searching bark crevices.

The male is crisply patterned in black and white, without the splashes of color found in many warbler plumages. The species breeds as far north as the Northwestern Territory of Canada and winters south as far as Peru. It prefers the canopy of large trees for its foraging, but it nests in leaf litter on the ground.

It was places such as Clark's Creek that I was targeting on my journey—gorgeous green spaces with singing warblers tucked away out of sight. From Clark's Creek, I headed north and west to another such place, Lake Bruin, an oxbow lake of the Mississippi lying about a mile north of the small agricultural town of Saint Joseph, Louisiana. Sixty-four-acre Lake Bruin State Park is tiny but picturesque, sandwiched between the lake and the great river. Here the rich black bottomland earth is spread as flat as a billiard table, waiting to be planted with soybeans, cotton, milo, or corn. Two things break the monotony of the flat land: to the east rises the massive grass-covered earthen levee constructed by the U.S. Army Corps of Engineers to hold the Mississippi within its banks, and a bit north, near Newellton, rise three fifteen-foot Balmoral Indian mounds, remnants of the Mound-Building Native American culture that produced curious earthworks throughout the eastern and central United States a couple of centuries before the arrival of Columbus. Together the structures are an intriguing historical echo: precolonial Native Americans made riverside earthworks that foreshadowed the great earthworks later constructed by the Corps.

The Native American mounds were built for religious, funeral, or cultural reasons, but the levee was constructed for flood control. The levee is the most prodigious human earthwork ever constructed, spanning thousands of miles. As Alan Lomax wrote: "The levee is unobtrusive, since its slope is green and gradual, but in fact it is immense—higher and longer than the Great Wall of China. . . . It was the principal human response to the titanic power of the great river."

Just outside the entrance to Lake Bruin State Park, a small gravel road took me to the levee crest. I drove carefully up to its flat top and found a well-maintained single-lane dirt road that followed the flat top of the berm. From atop the levee I gazed down to the muddy river, running high and flooding the maple-dominated riverbank forest just below me. From here, I appreciated just how big the levee actually is. In this flat land, its summit provides a view far and wide across a mix of river, swamp, forest, and rural agricultural landscape. I could see the church steeples of Saint Joseph a few miles to the south, and I finally understood what Don McLean was talking about in his ballad "American Pie"—a teenager could indeed bring a Chevy up here to hang out on a lazy Saturday afternoon or evening, drinking and passing the time.

Tall oaks and Sweetgums dominated the state park, a mecca for local bass fisherman because of its access to the blackwater oxbow lake that forms the park's western boundary. It was an ideal spot to set a tent for several days and gain my bearings in the interior bottomland delta country of the Deep South. A Barred Owl hooted in the afternoon daylight, and a Northern Mockingbird sang from afternoon into the night, illuminated by a full moon.

THE TENSAS AND THE U.S. ARMY CORPS OF ENGINEERS

In the morning, I rise to a cacophony of bird sound—Northern Cardinals, Tufted Titmice, Carolina Wrens, and Mockingbirds, all local resident species. On an early-morning bike ride around the small park's road network, I encounter flocks of Chipping Sparrows foraging on the ground in clearings among the campsites, readying themselves for their flight to breeding grounds in the Mid-Atlantic states or farther north. A flock of fifteen Indigo Buntings forages in an empty parking lot beside the water. White-throated Sparrows, perhaps on their way to the Adirondacks to nest, scuff in dried leaves under small thickets. Nesting Red-headed Woodpeckers chivvy in the oaks overhead.

I was in northeastern Louisiana to explore the forests and birdlife of the Tensas (pronounced "TEN-saw") River. But first I checked out nearby Saint Joseph, an antiquated low-country hamlet with an unusual feature: its little main street comes to an abrupt end where it meets the high levee of the Mississippi. It looks as if the great river has cut off the eastern part of the town. The levee was constructed, one presumes, to keep the rest of town from being gobbled up by the wandering river.

The levee rising above the eastern verge of Saint Joseph is part of a network of earthworks that seeks to keep the Mississippi within its historical course and to protect floodplain communities such as this one from catastrophic inundations. This earthen structure is about fifty feet high and about twice as wide at its base. More than 3,600 miles of levees have been constructed in the Mississippi drainage by the ever-busy Army Corps of Engineers, and one of the hydrological problems created by such systems is that the confinement of the river between two earthen barriers prevents the river from overflowing and dropping its silt onto floodplains during periods of high water. Instead, the silt sinks to the bottom of the main flow of the river and actually raises the river in its own course. This has most famously happened in the city of New Orleans, which now lies *below* the level of the flowing river. In fact, this has happened all along the Mississippi wherever there are levees; here at Lake Bruin, the river nearly breached the levee in 1991. Thus, decade after decade, the levees sow the seeds of their own destruction and generate a need for more and bigger fixes by the Corps. Because of its claims of economic necessity and the indisputable vulnerability of riverside communities, the Corps gets its way; congressionally authorized funds for its expensive engineering projects are spent in local districts, so there are many winners, who in turn become noisy boosters for the Corps' next projects. The losers are, of course, the general public, wild nature, and agriculture, which is deprived of all the fertile topsoil that, without the levees, would be spread back onto farm fields in the floods of spring.

Founded in 1802, the U.S. Army Corps of Engineers, with more than thirty-seven thousand civilian and military employees, is the most influential federal government agency overseeing the management of America's rivers, estuaries, and coastlines. The Corps was responsible for the construction of the Panama Canal, the Bonneville Dam, the Washington Monument, and, of course, the management of the Mississippi as a critical industrial waterway. The Corps owns and operates

609 dams, 257 navigation locks, 75 hydroelectric plants, 12,000 miles of navigation channels, and 926 harbors across North America. It dredges 255 million cubic yards of sedimented muck per annum and stores 329 million acre-feet of water in reservoirs. The Corps, then, following its mandate of improved commerce, flood control, and energy generation, has changed the face of America's natural waterways and conducted engineering experiments on some of the most important freshwater ecosystems in North America. To many Americans, the Corps is synonymous with environmentally questionable boondoggles. The levee system that failed in 2005 during Hurricane Katrina, flooding the city of New Orleans, epitomizes the overexpenditure and underperformance of this massive federal bureaucracy. The next big failure is expected to be the Old River Control Structure in central Louisiana, which currently prevents the Mississippi from changing course and following the Atchafalaya to the Gulf. That failure, when it happens, will dwarf the economic impact of Katrina.

ECHOES OF THE IVORY-BILL

My tent site at Lake Bruin was about a quarter mile from the big, grassy levee that hems in the west bank of the Mississippi. If I squinted, I could just make out the high green berm through the oaks. But if I drove west about fifteen miles, I would enter the watershed of the Tensas River, the last bastion of a near-mythical bird. Anyone who has read Arthur Cleveland Bent's *Life Histories of North American Woodpeckers* or James Tanner's *The Ivory-Billed Woodpecker* will recognize the name of the Tensas River, in the heart of delta country. The Tensas is where Tanner studied the last known breeding population of Ivory-billed Woodpeckers, America's largest and most strikingly patterned woodpecker, from 1937 to 1939. Being a woodpecker fanatic as a preteen, I'd first read Bent and Tanner in the early 1960s. The crisp monochrome images of the magnificent woodpeckers at their nest hole in a big Red Maple, photographed by Arthur Allen and published in the Bent and Tanner volumes, are among the most iconic photos in all of American ornithology.

Until the late 1930s, the Tensas was still dominated by grand forest and wooded swamps, with remnant populations of Black Bear, Cougar, and Red Wolf as well as the Ivory-bill. President Theodore Roosevelt had hunted swamp-dwelling Black Bear in the Tensas Bayou, just north of Tallulah, in 1907. Guided by Ben Lilly and Holt Collier, the president and his party decamped from their train at Stamboul, Louisiana, a far cry from a tourist destination at the time of their visit. Only a wilderness fanatic like Roosevelt would have had the urge to battle the briars and chiggers and ticks and Cottonmouths in search of a southern swamp bear. At that time, the only other reason to visit here was to search for Ivory-billed Woodpeckers, for this was the last place they lived in numbers. Roosevelt later wrote:

> The most notable birds and those which most interested me were the great Ivory-billed Woodpeckers. Of these I saw three, all of them in groves of giant cypress; their brilliant white bills contrasted finely with the black of their general plumage. They were noisy but wary, and they seemed to me to set off the wildness of the swamp as much as any of the beasts of the chase.

But the approach of World War II witnessed the decimation of the last tracts of old-growth hardwood forest in the Tensas, as well as the destruction of the last Ivory-bills by hunters who could not resist taking a shot at a big black-white-and-red woodpecker. Subsequently, commodity booms over the past half-century have led to the conversion of much of the remaining bottomland forest to monoculture cropland, and thus today the Tensas basin is a far cry from what Roosevelt saw when he toured this southern wilderness. The good news, though, is that the U.S. Fish and Wildlife Service has actively reforested a large section of the Tensas, and a number of the fertile river bends that once held Ivory-bills now support mature hardwood forest that grows grander every decade. The Ivory-bill is gone, but the Black Bears are back, and hundreds of other species prosper here, including some of my quest birds.

This wooded swampland supports ridge and swale hardwoods, growing on fine, dark sediments produced by wet-season flooding of the Tensas River, which is surprisingly small and unprepossessing. Hidden in the tall hardwoods, this river flows muddy in spring, with high mud banks on either side. It is not surprising, however, that the Tensas's grand forests arose in bottomlands distant from the main flow of the Mississippi. Along the Tensas, conditions favor timber growth without the catastrophic disturbance produced by the big river's major floods. The Tensas floods just enough to fertilize everything without regularly tearing it all down (and, of course, the Tensas is not leveed). I had seen the same thing in the jungles of New Guinea, where the finest stands of timber lie distant from the great river channels.

The Tensas River sinuously bends its way southward from the Tensas Bayou, just west of the town of Transylvania, a few miles from the Arkansas border. South of the Tensas National Wildlife Refuge, the river winds south until it meets the Black and Ouachita (pronounced "WOSH-i-taw") rivers, east of the town of Vidalia. The Tensas bottomlands include Louisiana's most famous hunting grounds for Wild Turkey and White-tailed Deer, and it was the height of turkey season as I tooled around the gravel and orange clay roads, dodging big, muddy potholes. Given how impressive the forests I saw there are, I could only try to imagine what they had been like before the big timber was cut in the 1930s. Giant swamp hardwoods must have darkened the sky with their broad, leafy canopies.

At the spiffy headquarters and visitor center of the Tensas River National Wildlife Refuge, I found I was standing on the river's Greenlea Bend, which in 1937 was home to a nesting pair of Ivory-bills that Tanner studied. I was impressed by the majestic second-growth forest of Nuttall Oak and Sweetgum. I met with Kelly Purkey and Tina Whitney, two refuge staffers, peppering them with all sorts of questions about the location of the best patches of big forest accessible by kayak. We formulated a plan that would allow me to see the best of this refuge's resurgent forests.

The next morning, armed with their annotated maps, I headed into the Tensas for some first-hand exploration. Dodging luxuriant tangles of Poison Ivy, I dropped my kayak into the café-au-lait waters of Big Lake, a widening of the Fool River, which flows ever so slowly into the Tensas at the head of McGill Bend. In the adjacent bottomland woods, the oak canopy showed dark, leafy greenery, and by the sounds emanating from the trees, it was clear that this was woodpecker heaven. Within a few short minutes I heard the strident notes of various Pileateds and Red-bellieds, a Hairy, and a Flicker. At one point farther along the river, three different Pileateds called out from three different directions, while a fourth one drummed. Prothonotary Warblers and Northern Parulas sang from the forest canopy, along with innumerable Carolina Wrens, those champion southern songsters. No wonder the Ivory-bill had loved it here.

A frog that I did not recognize gave a rapid monotone trill. Several alligators retreated and submerged themselves as I paddled down to meet the Tensas at McGill Bend, where the river is perhaps seventy-five feet wide and great for kayaking. That day the water was high, which meant I didn't have to deal with fallen trees and snags, all now safely submerged. The kayak made not a sound, and I heard only birds and frogs and toads—a superb natural morning chorus free of cars or jets or construction or ringing telephones. Here one can feel a bit of what Tanner must have felt when he searched these big woods for Ivory-bills in the late 1930s.

After four hours on the water, I was back in the car to wander the muddy, unpaved forest roads and extricate myself from the confines of the Tensas bottoms. I came upon Crystal Road, which did not even show up in the big red Delorme Atlas that sat open in my lap. One of several backroads that cuts through the grandest of the Tensas's forests, it is a heaven-sent transect through the wilderness. Here all the familiar songbirds declared their territories: Northern Cardinal, Carolina Wren, Tufted Titmouse, Prothonotary Warbler, and White-eyed Vireo.

Then I heard a bird I had been anxiously listening for—the elusive but vocal Swainson's Warbler, a denizen of these swamp forests, and another of my quest birds on its breeding territory. I was, in fact, now in prime Swainson's country. One of the plainest of wood warblers, the Swainson's looks a bit like a Worm-eating Warbler but lacks the black crown stripe. Instead, it features a dark eye-line, a chestnut cap, a buff-olive back, a pale breast with a slight yellow wash, and a white undertail. It makes up for its dowdy plumage with its powerful musical song, which rings through the swampy forest—*Seeyu-seeyu-sa-see-sa-yu!* This singing male had perhaps recently arrived from its winter haunts in Cuba or the Yucatán. This is a species that nests on the ground in thickets of cane within openings in mature swampy forest, and I had managed to find one of these prime patches of habitat. This was the only Swainson's Warbler I was to encounter on my travels except for the handheld bird shown to me by Keegan Tranquillo in Louisiana.

A big, fat Canebrake Rattlesnake coiled at the edge of the road. Its presence, and that of occasional Cottonmouths, had kept my kayak expedition a prudent one and would make any hikes in the woods a risky proposition. I had left my tall rubber boots back at camp, so my feet and legs were unprotected should I tread on a venomous snake while bushwhacking. A bit farther down the road, I came upon a handsome—and nonvenomous—Black Rat Snake as well as butterflies galore: Spicebush and Tiger Swallowtails, Red-spotted Purple, and Red Admiral, all wavering about on this spring day in the woods.

At midday, returning from the Tensas bottoms, I wandered through the impoverished neighborhoods of Tallulah, the nearest town, in search of a place to eat lunch. It turned out to be Louann's, on the town's outskirts, with a couple of tables of good ol' boys enjoying local delicacies: mustard greens, fried okra, hushpuppies, and frog legs. With my notebook and field guide, I probably looked out of place, but everybody was friendly, and the tasty food was well fried.

On the following morning, I was back in the Tensas to tour the forest interior of McGill Bend with refuge staffer Nathan Renick. We ATVed the seven miles around the curve of the river to get deep into the bend's forest, a mind-blowing ride—Renick jumped six-inch logs, traversed standing water three feet deep (which sloshed into the open cab), got partially stuck in deep mud, cut right through palmetto thickets, and did things I did not know could be done in a motorized vehicle. Renick, a bear of a man, carried a twelve-gauge shotgun for Wild Boar, a feral species that does serious damage to the forest's low vegetation. We parked in an understory of Saw Palmetto and bush-whacked for a couple of hours in search of wilderness goodies, Renick pointing out interesting features of the landscape and telling me one story after another about the Tensas.

There is nary a pine in the Tensas basin; instead McGill Bend and the contiguous refuge forest constitute a core preserve of about eight thousand acres of hardwoods—a big chunk of mature and road-less bottomland forest nestled within the larger refuge boundaries. The forest of McGill Bend is even-aged, all about seventy-five years old. We saw not a single giant tree here, except an old cypress down in the waters of the Tensas, but the forest was tall and impressive, made up of five species of oak: Overcup, Cherrybark, Water, Nuttall, and Willow. American and Cedarbark Elm, Sweet and Bitter Pecan, Sweetgum, Green Ash, Sassafras, Southern Hackberry, Honey and Black Locust, and Persimmon filled in the woods. There is no remnant old-growth forest here—the Chicago Mill and Lumber Company got it all, down to the last big timber tree—but, even though it is not very old, this is rich bottomland forest, where Ivory-bills lived a few generations back. We saw none of them, of course, but we did see bear tracks and photo-graphed a Copperhead snake, listening to forest song dominated by Carolina Wrens, Summer Tanagers, Great Crested Flycatchers, Tufted Titmice, and Prothonotary Warblers.

As I drove back to the campground on graveled Mill Road, an amazing profusion of butterflies greeted me, including thousands of Eastern Commas and Snout Butterflies. I had never seen such an aggregation before. Was this some sort of seasonal emergence? Farther on, in the back country, the GPS instructed me to make a left turn where there was nothing but regenerating forest; a formerly extant road must have been abandoned here where the refuge carried out large-scale reforestation to expand its bottomland forest. This was rewilding in action. Renick had told me the reforestation plantings included all seventeen dominant canopy tree species known from the area, and I wished I could return to see this forest seventy-five years hence.

Three decades after Roosevelt hunted bear in the Tensas, Arthur Allen journeyed here to photograph and record the last of the Ivory-bills, and his student James Tanner spent much of the following several years visiting the great forests of the Tensas to document the lives of the woodpeckers. Tanner's was the only field study ever conducted of this wonderful and tragic species; although he undertook fieldwork throughout the Southeast, he found Ivory-bills only here. During the latter part of his study, the Chicago Mill and Lumber Company was taking down the Tensas's great virgin stands of hardwood. In spite of the efforts of many parties, both state and national, to preserve the forest for the remnant population of Ivory-bills, the last of the big trees were removed by 1944. That year, artist Donald Eckelberry observed a lone female Ivory-bill in the forest—the last one that was ever seen in the Tensas.

A public announcement was made at a press conference on April 28, 2005, that the Ivory-billed Woodpecker had been rediscovered in southern Arkansas, a couple of hours' drive north of the Tensas. At that point, a brief video clip shot of a bird in flight near Bayou De View was the only evidence for the continued existence of the species. High-tech field surveys followed, undertaken by Cornell's Lab of Ornithology in partnership with the state and federal government, the National Audubon Society, and the Nature Conservancy, meant to generate information about the status of this rediscovered species. But time passed, and no additional evidence was obtained. Various independent scientists publicly expressed doubt about whether the original video even depicted an Ivory-bill. Perhaps it showed the widespread Pileated Woodpecker. People interested in the story then sorted themselves into "believers" and "doubters." More than a decade after that exciting press conference, there is no evidence that a population of this great woodpecker remains alive in Arkansas or any place else on earth. The ornithological world has moved on, a bit saddened, and most of us now relegate the Ivory-bill to the realm of the extinct. That said, I am confident that, with advances in molecular genetic techniques, a facsimile of an Ivory-bill could be engineered in a laboratory via technology that is probably one or two decades off. Whether people in white lab coats can re-create the Ivory-bill or instead will create a bird much akin to it, with many Ivory-bill genes, is not clear. One day, those manufactured birds could be reintroduced into some of the oldest and wildest bottomland forests of the Deep South. Suffice it to say, I am perfectly OK with a lab-constructed Ivory-bill being brought back to the Tensas and allowed to recolonize the forest that is now protected as a national wildlife refuge.

BOTTOMLAND WOOD WARBLERS AND FOREST LOSS

Mid- to late April in northeastern Louisiana is wood warbler paradise. A diverse mix of passage migrants and nesters were here, for the delta forests offer rich and productive breeding habitat for these diminutive insect-eaters. The breeders, in most cases, arrive first, before the passage migrants that are headed farther north. Logic might argue that the birds with farther to travel should depart earlier, since their trip will take longer. But what matters most is whether the breeding habitat is ready to receive the birds. At this time, it was still snowing in northern Ontario, where many of the passage migrants nest.

My field guide maps showed that these low-country forests were breeding habitat for ten species of wood warblers. Virtually every patch of forest rang out with the territorial songs of at least one or two species. That said, most are habitat specialists, and I had to work hard to find any of them. The density of these birds on territory was quite low—perhaps due to the annual mortality during the long wintering period.

The passage migrants that stop over in the forests of the Tensas probably exceed twenty species, yet in my several days in the Tensas, I recorded only a handful. Tennessee and Myrtle Warblers were common, but I found few other species: only the migrant Magnolia, Chestnut-sided, and Yellow Warblers. When there are eighty thousand acres of forest, passage migrants tend to get lost in the woods. Only a few were singing at this stage, which made them even more difficult to locate. What is bad news for the birdwatcher is good news for the birds, though. The Tensas has abundant habitat for wood warblers—both those staying to breed and those on the way through—in spite of large-scale loss of forest in the Mississippi Delta over the past century.

Nonetheless, anyone who has driven the back roads of the delta region of Louisiana and Mississippi can see how much forest has vanished from this fertile agricultural region. In April, one encounters vast planar fields entirely bare of vegetation, waiting for their spring crops of cotton, milo, corn, and soybeans. Tall forest stands in the distance are usually in wet bottoms that can't support agriculture because of the annual inundation from spring floodwaters.

Farming has been prominent in the delta for nearly two centuries, and the Mound-Building Native American cultures who lived here prior to colonial settlement probably cultivated extensive areas as well, given the fertility of the region's deep, black, loamy soils. In fact, there probably have been repeated cycles of deforestation and regeneration, but it is safe to assume that this last pulse of clearance has been the broadest, in part because of the evolution of agricultural practices and the size of modern business operations. To maximize scale, agribusinesses have sought to generate wall-to-wall row crops. Hedgerows and bordering tree lines have been removed, and edge vegetation that can support bird populations of many kinds has been much reduced.

A look at this area using Google Maps in satellite mode shows a large lens of prime agricultural delta land extending from southern

Louisiana north-northeast to Cairo, Illinois, where the Ohio River meets the Mississippi. This large deforested area includes central Louisiana, eastern Arkansas, western Mississippi, southeastern Missouri, and eastern Tennessee. This is low country that was periodically an inland seaway, and it is flat and alluvial because of the absence of hardrock outcrops that could impinge the flow of the great river. The giant lens of fertile silt was created over time by the relentless back-and-forth movement of the Mississippi as it swept across the landscape while in flood, razing forests, moving sediments, and building floodplains.

A glance at the satellite imagery also reveals linear swathes of forest lands in a number of river bottoms, including that of the Mississippi. These serve as important forest reserves for millions of migrating and nesting Neotropical songbirds. One also notices that to the west and east of the lens of agricultural land are extensive hilly areas cloaked in a mix of hardwoods and pinelands. These, too, offer habitat for migrating and breeding songbirds. So the news is mixed: some good, and some bad. Certainly, the documented songbird declines are troubling, but perhaps the pendulum of loss has slowed; time will tell. Needless to say, we need to encourage state and federal agencies and private landowners to restore more forests on devegetated lands, and we need to encourage farmers and agribusinesses to adopt more forest- and hedgerow-friendly practices. Regreening bare land is also essential if we hope to address the long-term threat of climate change.

Actually, ornithologists are not the only ones who appreciate the presence of extensive bottomland forest in the Deep South. Our closest allies are people who are, like Nathan Renick, avid sport hunters of Wild Turkey and White-tailed Deer: they, too, want to preserve and regreen forest lands. Aside from fishing and rooting for a favorite SEC football team, heading into springtime or autumn woods to hunt is a favorite pastime in the low country of Louisiana, Mississippi, and Arkansas. Hunters' enthusiasm reaches a religious pitch with respect to turkey, one of the most challenging quarries known. In addition, hunting makes a substantial seasonal contribution to the rural economies of many small towns in these parts.

A HUNT FOR BIG TREES

Small patches of ancient timber do remain in the Mississippi Delta, but these are few and far between. As I have found, the Tensas, sadly, holds none at all. So I head to a remnant old-growth patch in northeastern Mississippi, in Delta National Forest, which holds a mix of forest and nonforest habitats and is surrounded by row-crop agriculture, as one expects in the delta. From Lake Bruin, it takes me a bit more than two hours to find the gravel road to Sweetgum Research Natural Area. I set up my tent in its parking lot, and I head into the forest to look for big trees.

This area—about forty acres tucked amid sixty thousand acres of national forest—features ancient Sweetgums scattered throughout a single old-growth plot. The plot is species-rich, boasting oaks, Prickly and Green Ash, Honey Locust, and Southern Hackberry. Many of the Sweetgums top 130 feet in height, and several have trunks exceeding four feet in diameter. The largest Sweetgum I saw, 135 feet tall, was more than five feet in diameter at breast height. No doubt, these four-hundred-year-old trees hosted foraging Ivory-billed Woodpeckers at some time in their long lives. It was enthralling to stand beside the forest giants and think back two centuries, when they were already postmature canopy trees. Moreover, it was sad to think that across the length and breadth of this immense tract of national forest, only a handful of old-growth trees were preserved for posterity. What were foresters thinking when they did not set aside larger samples of the original ancient forest? Should such strategic set-asides not be a mandate of any "national" forest? Those overseeing the government management of our western forest lands, which still harbor ancient tracts, need to take note.

I spent a day and a half birding and naturizing in Delta National Forest. The access road cut through a fine stand of secondary forest about a half-mile in extent, through which I biked to a wooded swamp-land just to the north. A very wary Red Fox—not one of those tame residential creatures I knew from home—skittered across the road in front of me. At the forest's wetland, Mississippi Kites played about in the sky over the prime swamp habitat. Back at the parking-lot campsite, by the stand of old trees, I heard a Prothonotary Warbler, Kentucky Warbler, Acadian Flycatcher, Veery, and Wood Thrush. I had expected to hear many more forest-dwelling birds, but instead the soundscape was dominated by the voices of lots of nonforest species, such as the Carolina Wren, Great Crested Flycatcher, White-eyed Vireo, and Mourning Dove. And, just as at every forest edge in the early morning, flocks of Indigo Buntings were coming up out of the roadside grass. Huge numbers were moving northward to their breeding territories in the middle of the country—I could not escape the little blue songbirds.

A male American Redstart, singing on territory, was my tenth quest species. It darted about in the understory at the edge of the forest, chasing down winged insects among the low shrubbery. Breeding from southern Louisiana north to Labrador and the Yukon, this is one of the most familiar forest-dwelling warblers in North America, beloved for its confiding and active demeanor, its bright and upbeat rapid song of slurs and chips, and the male's plumage of black, orange-red, and white—a bit reminiscent of an undersized Baltimore Oriole. I clearly recalled that I'd seen the species for the first time in the early 1960s in the tall woods on Bill Johnson's farm, north of Baltimore. I'd whooped with delight when the male warbler danced in front of me, showing off his bright oriolelike colors.

Foraging in a shrub at the forest edge near my tent was another surprise: a Ruby-crowned Kinglet. I had forgotten that they wintered so far south. In fact, this tiny, high-energy species was a sort of fellow traveler on its way to the far north: it, too, was headed to Ontario, where it would prove to be one of the commonplace species in my little-known destination, the upper reaches of Kenora District in northern Ontario. The kinglet is not a wood warbler, but it is certainly reminiscent of one, with its olive plumage, white wingbars, and pale eye-ring. In addition, it's as hyperactive as a warbler. Weighing a mere quarter of an ounce, this kinglet is North America's second-smallest songbird—only its cousin, the Golden-crowned Kinglet, is smaller. Although they were not as eye-catching as the Indigo Bunting, swarms of kinglets also would be leading me northward, and welcoming me at the apex of my journey with their bright, bubbling songs in a few weeks' time.

It was almost time for me to leave the vast alluvial delta of the Mississippi. The delta continues to support bottomland forest in the annually flooding low country of the many tributaries of the Mississippi, including the Tensas, Atchafalaya, Red, Black, White, Ouachita, and the Saint Francis—good news for migrant songbirds. The flat alluvial country penetrates northward into Missouri and Illinois, near the confluence of the Ohio and the Mississippi, and the delta ends where the Ozarks rise up in eastern Missouri and the Shawnee Hills break up the landscape of southern Illinois. Bluffs and hills announce the inevitable transition from the black-soil South to the upland North, where the Mississippi winds between tall, rocky banks vegetated with handsome and well-drained upland hardwood forest. Before I headed to the North, however, I needed to explore two additional Deep South songbird habitats: the piney woods of Texas and Arkansas, and the associated cypress swamps in those two states and in Missouri and southernmost Illinois.

FIVE

Piney Woods and Cypress Swamps

Late April 2015

Out of the luminous mist in which the trees showed like ghosts I heard the forgotten voice of yesteryear—once and silence, again and silence, and again—at regular intervals from a grove of trees, the little buzzing, ascending wisp of song that the parula warbler gives off.

—LOUIS HALLE, *Spring in Washington*

The southern pinelands—or piney woods—cloak tens of millions of acres in the South. Migrant songbirds breed here, as do several rare and localized resident bird species, all prime incentives for me to visit the area. Pines grow in abundance in the uplands, but down in the flooded swamplands, the cypress flourishes.

Much of eastern Texas, western Louisiana, and southern Arkansas hold piney woods, and additional southern pinelands range from central Mississippi and northern Florida northeast to southernmost Virginia. Whereas the delta, with its deep black soils, supports row-crop agriculture, the pinelands' sandy soil generates timber and pulpwood. As a result, after they are logged, these areas are quickly recycled back into monocultures of production pine forest. The good news is that the pinelands remain in a forested state. The bad news is much of this pine-forest acreage is overmanaged for timber and pulp, with relatively little left in its natural condition. Nearly gone are the magnificent and parklike Longleaf Pine savannas of the South, where tall pines once towered over sun-dappled grasslands. The open nature of those woodlands was maintained by periodic wildfire, but over the decades, as the pines were harvested for all sorts of construction tasks, from shipbuilding to flooring, wildfire was suppressed. Now less than 1 percent of that fire-dependent habitat remains.

Cypress, meanwhile, prospers in the standing blackwater swamps of oxbows and backwaters along the many tributaries of the Mississippi, from Louisiana north to southern Missouri and southernmost Illinois. Just as the piney woods are favored habitat for an array of songbird migrants, the cypress swamps attract their own songbird specialists. I searched for these birds with the help of various local naturalists as I traipsed from Texas and Arkansas to Missouri and Illinois.

ORIENTATION: THE INTERNAL COMPASS

As I moved about from site to site, I sought to get a sense of this large inland region. The wood warblers I followed did the same thing, not using maps and GPS as I did but instead employing specialized

OPPOSITE: Red-cockaded Woodpecker

sensory faculties to orient themselves with respect to where they were and where they needed to go to arrive safely at their breeding grounds. Orientation is pretty simple in theory—it is the ability to distinguish north from south, and east from west. It essentially means that you have access to a compass. There is no doubt that migrant songbirds have the capacity to orient themselves properly with regard to compass direction. The mystery, of course, is how they manage to achieve this without having access to an actual compass.

Yearling songbirds in their northern breeding habitat first learn to orient in the late summer, when they must head south to their winter habitat for the first time. First they experience an innate *urge* to disperse from their natal territory, based on the birds' internal annual calendar. This calendar is kept in calibration by external astronomical cues, including day length and the path traced by the sun across the sky each day. The birds also possess the innate ability to determine *direction*—an internal compass. What is the nature of this compass? We now know that birds possess several tools to orient themselves. The first is the sun. By tracking the sun's path across the sky, a bird, with knowledge of the season of the year and the time of day, can orient itself to a proper compass direction. Second, birds can also detect the plane of polarized light in the sky—and thus the position of the sun—even after sunset or on cloudy days.

That said, many migratory songbirds migrate at night, well after the sun has set. What then? Songbirds use the starry night sky to determine compass direction. In their first summer of life, birds learn how to detect the rotation of the constellations around the North Star to determine north. They subsequently can use the pattern of the constellations to locate the North Star, from which they can orient themselves without having to detect stellar rotation.

But there's more. Birds possess an internal magnetic compass: the ability to detect the lines of the earth's magnetic field, which allows them to determine magnetic north. Birds and many other animals have small particles of magnetized iron (magnetite) in their bodies that apparently aid in this ability to orient themselves. In pigeons, the magnetite is located between the brain and the skull. Other research indicates that birds are able to generate magnetically sensitive internal chemical reactions that can serve a compass function.

Thus birds appear to have a minimum of four separate tools that provide compass orientation and guide their travels. They rely upon these multiple capacities because, just like a car with both a parking brake and a pedal brake, it is good to have backup systems for critical tools that one depends on in life-or-death situations.

Yet birds have still another trick up their wings. They can tell *where* they are on earth, not merely how to get there. In a later section, we'll discuss that internal GPS capacity.

CYPRESS AND BIGFOOT

My first stopover is Caddo Lake State Park near Karnack, in northeastern Texas, not far from the border with Louisiana and Oklahoma. I'm here to spend five nights at cypress-filled, 25,400-acre Caddo Lake, which straddles the Texas-Louisiana line. Karnack, so small that it has no downtown, just a couple of intersections where convenience stores have settled, attracts visitors both to the state park and to Caddo Lake National Wildlife Refuge, where I plan to hang out with a university group doing a natural history field survey.

I set up camp in a corner of the park that I have all to myself: a spot in tall bottomland forest at the edge of a blackwater arm of the lake. Cypress trees rise from the water on one side of the tent, and great oaks on terra firma provide shade from the scorching Texas sun on the other. I'm deep in the woods near where Big Cypress Bayou flows into the lake. Soft green light filters onto the picnic table, and a male Prothonotary Warbler sings in a nearby tree. I eat my lunch watching the orange-tinted songster declare his territory in his unmusical, repetitive lisp.

Caddo Lake State Park was created in 1933, when Lady Bird Johnson's father and several other local landowners contributed the property needed for the reserve. Using the National Park Service's "natural design style," the Civilian Conservation Corps (CCC) constructed the park roads and buildings, completing the project in 1937, at the height of the Great Depression. A number of the rustic CCC buildings, made of timber and stone sourced locally, still stand in the park today.

Caddo Lake is famous for a few reasons. First, it has been identified as a wetland of international importance through the Ramsar wetlands treaty (which designates globally significant wetlands around the world to encourage their conservation). Second, it is a favored weekend hangout for Eagles drummer Don Henley, who grew up in Linden, a bit north of Karnack. Henley founded the Caddo Lake Institute to foster conservation of the lake's ecosystem and to assist with reintroduction of the endangered American Paddlefish to the lake's waters. He is an example of a private citizen, albeit a well-known one, who has done more than his share to support conservation of habitat and endangered species because of his love for a verdant corner of the world that happens to be in his childhood backyard. And because northeastern Texas is rich in fishermen, Caddo Lake's Large-mouthed Bass, White Bass, Crappie, and Sunfish are other reasons for its popularity. The lake is perfect for kayaking and birding as well.

Caddo is a large blackwater lake adorned with huge stands of mature Bald Cypress, an antediluvian swamp conifer of the Deep South. The big trees, festooned with Spanish Moss, are its most remarkable and otherworldly feature, and, of course, they capture the attention of every visitor. Ringing the shallows, this cypress is a tree apart. A strange conifer that drops its leaves in autumn, it has knobby "knees" (function unknown) that rise out of the dark water like woody stalagmites. The tallest cypresses approach 150 feet in height, and the thickest trunks exceed sixteen feet in diameter. The most ancient of these cypresses is more than fifteen hundred years old. The wood of old-growth cypress is valued for its resistance to rot, and thus much was logged out by the early twentieth century.

Gazing at its ancient-looking cypresses and black water, one could imagine that Caddo Lake is a geologically ancient hydrological feature. Apparently not. The lake formed when Big Cypress Bayou was dammed by a landslide set off by the massive New Madrid earthquake of 1811. Thus Caddo Lake is a recent creation. In fact, geologists tell us that, in general, lakes are ephemeral features on the landscape, so perhaps this should not be a surprise.

Caddo's unearthly looks have given rise to still another reason for its fame (one that I speculated about at my campsite). There have been rumors of hundreds of sightings of Bigfoot, otherwise known as Sasquatch, in and around Caddo Lake since 1965, which is perhaps why one shoreline community sports the rather remarkable name of Uncertain. In 2015, the Animal Planet channel sent a film crew to do an episode of *Finding Bigfoot* that featured the obligatory nighttime field search for the creature in the Caddo Lake area. (Why do these searches always take place at nighttime, with night-vision head gear? It would be easier to locate the creature by setting out a few dozen camera traps along well-worn game paths.) The search team was unable to capture Bigfoot on film, of course, but they did interview local residents who claimed to have glimpsed the big primate, a species probably considerably less abundant than the endangered American Paddlefish.

The paddlefish is actually the strangest creature that we know for sure lives here. A primitive ray-finned fish related to the sturgeons, it has a prominent, long, paddle-shaped snout and can boast relatives dating back 300 million years. Over a paddlefish's thirty-year lifespan, it can reach a length of five feet and weigh as much as 150 pounds. The declining species inhabits the Mississippi basin and is primarily a filter feeder. Its population has dropped (as is the rule rather than the exception for large fish) as a product of overfishing, habitat destruction, pollution, and poaching for the fish's caviar, which is an inexpensive substitute for Beluga Sturgeon caviar from the Caspian Sea. Henley's Caddo Lake Institute has been working to reestablish American Paddlefish in Caddo Lake in partnership with the Texas Parks and Wildlife Department and the U.S. Fish and Wildlife Service. These groups have worked with the Nature Conservancy and U.S. Army Corps of Engineers to ensure that water flow from upstream is sufficient to provide suitable habitat for this odd and ancient fish. Time will show whether their important experiment is a success. At Caddo Lake State Park, I saw plenty of wildlife but not a single paddlefish.

STATE PARKS: FORGOTTEN GEMS

State parks are America's overlooked jewels. Most people know them as good places to camp, birdwatch, and fish, and typically state parks have infrastructure that makes staying at one a pleasure. At Caddo Lake State Park, for example, I camped in a tent in the forest, and yet each morning I could take a hot shower and shave in a spanking-clean bathhouse. Yet sometimes these parks—and there are 6,600 of them across the country—get a bad rap because they are not very wild or not terribly special biologically. But their glory is that they are local, accessible, and comfortable, and in many instances they offer the best of regional nature to visitors, many of whom come from nearby. Not to be confused with national parks, state parks' main purpose is to offer local residents a place to get away on a weekend or a Labor Day holiday. The thousands of state parks across the United States feature lakes, wetlands, rivers, forests, and mountains that offer recreation for us all—at close range and at minimal cost. Their success is measured by the fact that they receive nearly three times as many visits per annum as national parks. For most citizens, state parks are where they first learn about and encounter nature in a welcoming setting.

The second morning at Caddo Lake State Park, I noted a public bird walk was offered by park naturalist Mia Brown, and I joined the group of about ten participants in an hour-long introductory course in field ornithology. None besides Brown and me had ever been bird-watching, and she was able to introduce the newcomers to birding in a welcoming way. She had binoculars and field guides in the back of her truck, and the group was very enthusiastic as we looked for birds from a long dock projecting over the lake. Barn Swallows, an Eastern Phoebe, a Green Heron, a Great Egret, a male Prothonotary Warbler, and a male Summer Tanager delighted the group. Back at the truck, Brown took a small roll of paper towel off the front seat. She carefully unrolled it and showed the group a freshly killed Magnolia Warbler that, she explained, had struck her office window and fallen dead upon the sill. The bird was in perfect condition, and the children in the group were moved by its fragile beauty. Brown noted that windows are major bird killers: the birds see a reflection of green vegetation, attempt to fly through the space, and hit the pane. Her take-home point was that the human environment poses threats to birds.

That evening, Brown would offer an owl walk, and later in the week she'd hold programs on alligators and on the history of the Civilian Conservation Corps. Through thousands of such field programs with visitors, state parks enrich the experience of weekend visitors in important ways. I tip my birder's cap to Brown, and to all the staff who make state parks work.

HERP HUNTING

One creature I hoped to encounter on my journey north up the Mississippi drainage was not a bird at all, but a very large and frightening-looking aquatic denizen of the river's tributaries, swamps, and wetlands—an Alligator Snapping Turtle. The biggest individuals can weigh 250 pounds. As I was not a herpetologist, I did not know how to find one of these turtles, but Brown told me that a university group was spending the weekend surveying the herpetofauna at the adjacent national wildlife refuge. Perhaps they could show me an alligator snapper.

I tracked the group down at their makeshift encampment in a large utility garage in the center of Caddo Lake National Wildlife Refuge. There, herpetologist Rich Kazmaier and six of his students from the Department of Life, Earth, and Environmental Sciences at West Texas A&M University were busy photographing snakes, turtles, and frogs they had collected the night before. They were happy to include an extra participant, so I dove right into macrophotography of their diverse array of creatures, which they'd soon release: an Eastern Hognosed Snake, a Rough Green Snake, a Diamond-backed Watersnake, a Rough Earth Snake, a Green Treefrog, a Gray Treefrog, a Spring Peeper, and a Red-eared Slider.

Kazmaier told me that the Caddo Lake ecosystem is one of the richest in the Deep South for snakes, turtles, and frogs, which is why, each spring, he and his students drive nine hours from Canyon, Texas, to come here. I myself was already seeing creatures I didn't even know existed. While we photographed an adorable and cooperative Milk Snake, Kazmaier told me that he had netted a hundred-pound Alligator Snapping Turtle the previous year and that he hoped his traps would produce a similar behemoth this year. I crossed my fingers.

The team then ventured out for much of the night (without me), searching every likely spot for intriguing specimens. The next morning, I rejoined them at the refuge's shoreline access to Caddo Lake; I wanted to be present when the herp hunters checked the traps they'd set in the depths of the lake. The first creature they brought in was a Mississippi Green Watersnake, a specialty of the area but rarely seen. The nasty-tempered serpent repeatedly sank its teeth into the arm of one very stoic graduate student. Next they brought in a truly bizarre creature called a Three-toed Amphiuma, an eel-like aquatic salamander with vestigial limbs, no eyelids, and no tongue. The creature was about eighteen inches long, dark olive, and smooth-skinned. Another underwater trap held a two-foot-long Spotted Gar. But no luck on the Alligator Snapping Turtle.

We returned to the team's base to look at the terrestrial catch from the night before. Because they knew I wanted to photograph a Cottonmouth snake, they had brought one back for me (this species is so common and easy to identify that they don't usually bother to catch it during their nightly surveys). The snake was quite docile, so I was able to get very close to photograph it, along with a DeKay's Brownsnake and a Cajun Chorus Frog.

Although the herpetofauna is diverse in this part of the world, it does not fare well. Turtles crossing roads in spring get run over by cars, and nearly every snake that enters someone's backyard meets a hasty end. And then there are the infamous rattlesnake roundups, such as the one in Sweetwater, Texas. These roundups are annual events in which rural communities collect and dispatch every rattler

Cottonmouth snake

they can collect in a single long day of hunting. The objective is to rid the local environment of a venomous reptile, but in fact, these shy and long-lived creatures have a low reproductive capacity and are wiped out from areas where they are heavily collected. Most people, even those who love birds and other animals, do not give a snakes a break, and that's a shame.

Some youngsters have an affinity for nature and for natural history fieldwork, and the group Kazmaier taught was composed of just such kids. There was not a slacker in the bunch. Spending time with the group, I witnessed excitement and passion for nature among the students, who already knew a lot about local herpetofauna, birdlife, and plant life. And of course, their teacher was a true expert, with encyclopedic knowledge; Kazmaier loved both nature and sharing his knowledge with his students (a nine-hour drive is well beyond most professors' call of duty), and I am confident that at least several of the students will make careers in wildlife biology. These sorts of intense field experiences can transform the lives of students and encourage the growth of natural history field study in the United States. The classroom component is important, but it is in the field where students fall in love with their prospective profession. And these young knowledge keepers become leaders in the next generation, populating local, state, and federal agencies in charge of wildlife and parks and passing on their own knowledge and enthusiasm to yet another generation.

KAYAKING CADDO LAKE

In the afternoon, after my last visit with Kazmaier and his students, I drove south from Karnack on Highway 43 to look at the T. J. Taylor estate, where Lady Bird Johnson was raised by her father, a local "big man." The large white house sits above the road, handsome and substantial for the area but nothing out of the ordinary. Taylor, a widower, was smart and financially successful but also a bully. As a child, Lady Bird suffered with her unsympathetic father, and her marriage to LBJ was likewise difficult. But later in life she became something of a national saint. Her love of wildflowers and her campaign to beautify the nation's roadsides created a legacy that persists to this day. Lady Bird's flowers now bloom in glorious profusion in various places along Texas's highways, and as I drove I reflected on the power of individuals to change the world for the better. The accomplishments of Don Henley, Rich Kazmaier, and Lady Bird Johnson are examples to us all.

I spent my last dawn in Texas kayaking among the blackwater cypress in a narrow arm of Caddo Lake named Carter's Shute. A cold front had passed through, bringing blue skies and crisp temperatures. In places the fragrance of honeysuckle was overpowering. At the boat landing, I heard the distant, low, evocative drumming of a Pileated Woodpecker in the cypress. Low mist spread across the still, dark water as I followed a marked route through a flooded stand of the trees, draped with Spanish Moss. An Anhinga soared overhead. A Red-shouldered Hawk cried out in the distance. In a side channel, a Barred Owl perched low on a cypress stub. I quietly drifted toward it in my kayak as it mutely watched me. I snapped pictures until I got too close to use my long lens. I saw not another soul on my two-hour circuit. It was utterly peaceful.

As the sun rose, birds started to sing—first North Cardinal and Carolina Wren, and then Blue-gray Gnatcatcher, Prothonotary Warbler, Northern Parula, and—as an addition to my quest list—Yellow-throated Warbler, singing from the moss-draped canopy of a cypress. I had seen the species earlier on the trip, but only as a migrant. Here the species was breeding in the cypress swamp. Common in the Deep South and wintering in Florida, the Caribbean, and Central America, this crisply patterned bird has a big yellow throat patch, black and white markings on the face and flanks, white wing-bars, and a dove-gray cap and back. It has an unusually long and narrow bill, which it uses like a pair of fine tweezers to capture creeping insects in treetops. Its syrupy song is a descending series of slurs that abruptly ends with a *sweeet* note. Given the bird's canopy-dwelling habit, the song is critical to locating the species, which often nests in hanging bunches of Spanish Moss like those festooning the great cypresses of Carter's Shute.

BIG CORPORATIONS AND THE ENVIRONMENT

Late at night, a powerful thunderstorm passes over my campsite. Lightning strikes nearby, the wind blows, rain falls in buckets, and large branches tumble out of the canopy. I lie in my tent waiting for a branch to come down onto my tent, which once happened to students of mine in New Guinea (they survived, with some broken ribs). Perhaps now it is my turn. But luck is with me tonight.

My next stop was Crossett, Arkansas, about three hours northeast of Karnack, where I planned to shoot an environmental video with the Georgia-Pacific company and participate in GP's annual Water Ways Festival. The skyline of Crossett, deep in the piney woods of south-eastern Arkansas, is dominated by the large GP mill, which produces paper towels and toilet paper. West of town, I started to set up my tent beside the cypress-filled Ouachita River bayou, but before I'd done much, Terry and Sheryl, the park managers, came by in a golf cart to warn me that the bayou's water was supposed to rise twelve inches that night. They instructed me to move my tent to an upslope campsite.

At dusk, a neighbor at an adjacent site was busily cleaning several huge fish at his picnic table. I asked him the fish's name and how he'd caught them. In his eighties, the fisherman was friendly and had a sense of humor, like everyone else I met in the South. He happily said he was long retired but, to keep busy, worked as a freelance commercial fisherman netting Black Buffalo fish from the Ouachita River. The Black Buffalo can grow to nearly a hundred pounds, and its flaky white flesh, which he informally sold to locals in Crossett, is highly prized. The fish he'd caught that day each weighed more than thirty pounds and yielded a lot of fillet meat. This entrepreneur was making a killing selling them in town, he said; he was tickled to be able to do this work in his sunset years, as it made him feel useful.

As it got dark, a Chuck-will's-widow, the southern piney woods cousin of the more familiar Whip-poor-will, sang in a pine not far from my campsite. This southern night bird, which I'd heard sing only a handful of times, is not uncommon in the piney woods, but it is very difficult to see, as it roosts in a hidden perch during the day.

Early the next morning I birded the tall pines of the campground as I waited for the GP crew to pick me up, and I was rewarded with the highlight of another male Yellow-throated Warbler, this time singing from a tall pine. It would be a cool, sunny day for shooting the educational video; I worked in front of the camera with GP's wildlife program manager as we toured GP's wildlands, extensive local production forests of Loblolly Pine owned and managed by GP partners, and pine stands within nearby Felsenthal National Wildlife Refuge that support colonies of the endangered Red-cockaded Woodpecker. Passage migrants sang throughout the woods (a singing Swainson's Thrush was the high point), as did Neotropical migrants on their breeding habitat, including White-eyed Vireo, Northern Parula, Painted Bunting, and Pine Warbler—yet another quest warbler for me (number 12).

The Pine Warbler is a strict inhabitant of stands of pine during both the breeding season and in winter; to locate the bird, finding a good stand of pines is a prerequisite. It will inhabit just about any pure pine stand, even those planted as plantation monocultures. This warbler does not get much respect because it is rather dully plumed: mostly plain olive, with white wing-bars and undertail and a drab olive-yellow throat and breast, with some obscure streaking. In addition, it mainly winters in the vast pinelands of the Southeast, so we don't find this species winging over the Gulf of Mexico twice a year. As a breeder, the Pine Warbler ranges northward to southern Canada, so the species is widespread in the East. Perhaps the best thing about this bird is its song: a very sweet trilled series, invariably given from the canopy of a tall pine.

The following morning, I made my way to the Crossland Zoo, situated in a wooded section of Crossett City Park, and set up a "Spring Migration" table at the seventh annual Water Ways Festival, hosted by GP and focusing on the importance of water conservation and wise water management. I had been invited to host the station for groups of elementary school students, who were the event's target audience. Other institutions hosting visiting stations included the U.S. Fish and Wildlife Service, the University of Arkansas Division of Agriculture, the Boy Scouts of America DeSoto Council, the Crossett Centennial Garden Club, Georgia-Pacific, the Crossett Fire Department, the Crossett Rescue Unit, and the Crossett Zoo. Groups of fourth-graders visited each station for fifteen minutes to learn various aspects of water use and water conservation.

I hosted ten groups over the day, taking each batch of students on a short bird walk and talking about spring songbird migration. I'd worried that I might not find birds to show to the fourth-graders in a city park, but to the sharp-eyed, the woods were fairly birdy that day: Mississippi Kite, Gray-cheeked and Swainson's Thrush, Yellow-throated Vireo, and Orchard Oriole were in evidence, plus a Tennessee

Warbler in song and—best of all—a very cooperative Yellow-bellied Sapsucker that was managing a sap resource at a small elm tree near my station. The sapsucker tended its neat rows of drill holes about three feet above the ground, a perfect position for the students, many of whom had never seen a woodpecker. The children quietly approached to within about fifteen feet of the tree to take a good look at the foraging bird. Remarkably, despite the crowds and noise, the sapsucker remained at its elm from morning until the end of the festival in the afternoon. I told the students how the bird managed a water resource that provided it with sugars and other nutrients, and how it ate various insects attracted to the sap.

This sapsucker winters here in Arkansas, and this one would soon make its way north to its breeding territory in the Great North Woods. In fact, I wondered if I'd see this very same bird in late June in Ontario, when the forests ring with the cadenced drumming and squealing calls of the birds. The species is distinguished especially by its habit of drilling rows of small sap wells in certain favored trees, an unusual foraging specialization known only among the genus *Sphyrapicus*, the sapsuckers, which comprise four species that nest in the United States and Canada. Their favorite sap-producing trees are birches and maples (think maple syrup). Ruby-throated Hummingbirds appreciate the work of the sapsuckers and forage for sap at these manufactured resources, as do Cape May Warblers, bats, and even porcupines. The Yellow-bellied Sapsucker is the only completely migratory woodpecker; like Neotropical songbird migrants, it entirely departs its breeding range in the winter. It nests in the northern United States and Canada and migrates to a geographically distinct wintering range in the southern United States, Mexico, and Central America. I looked forward to making contact with the species on its breeding ground in the young mixed forests of Ontario; I grew up with these birds in the Adirondacks, and whenever I see a sapsucker now, it reminds me of summer in the North Woods.

It might be surprising, but one sponsor of my spring field trip was Georgia-Pacific, a major wood-products corporation, through a grant to the American Bird Conservancy. Although corporations are a part of America's wealth and its gigantic economy, big resource-extracting industries have done plenty of harm to the country's natural environment. Just think of the Chicago Mill and Lumber Company, which took out every last bit of old-growth forest in the Tensas basin. Big companies cleared ancient forests across North America at a prodigious pace between the 1840s and the 1940s—it boggles the mind to think of the billions of acres of ancient forest cut down during that hundred-year stretch. Yes, the timber was harvested to build American homes and other infrastructure, but the timber companies rarely left even an acre of original forest standing. Harvesting operations sought to maximize immediate yield, not long-term sustainability of the harvest, and there is no question that their focus on extraction changed the face of the eastern half of the United States forever. Only now is mature forest returning to some of the logged-out sites, and climate change may prevent the full and proper return of the old-growth forests that predominated across the East prior to the Civil War.

So, it is easy to speak negatively about America's extractive industries and their long-term record. But going forward, it is worthwhile considering how to effect change for the better, and most industries and corporations today are seeking to balance their negative environmental impacts (which continue) with their beneficial environmental offsets (which are more and more commonplace). For instance, GP partners with the Wildlife Habitat Council's Conservation Certification Program to improve stewardship over the corporation's extensive properties to benefit wildlife and nature. The interventions that GP carries out are audited by the Wildlife Habitat Council, and the council publicly recognizes outstanding efforts on behalf of nature conservation. Recognition of GP's conservation achievements then encourages other corporations to carry out similar beneficial activities for nature. Given the wealth of most of American corporations, it is rather straightforward for their leadership to commit to investing a percentage of the annual corporate budget in environmental good works. These can take place on the campus of the corporation (as is the case with Conservation Certification), or the corporation can make investments elsewhere. Amoco Production Company, as we saw in chapter 3, donated the oak woodlands and rookery pond on High Island, Texas, to the Houston Audubon Society, thus performing a permanent good—land conservation—that offsets the company's negative impact elsewhere.

The best results on this front tend to arise from partnerships between big corporations and environmental organizations. For instance, Walmart has worked closely with Conservation International to make its supply chain more efficient, thereby substantially reducing its use of gasoline and diesel fuel. This has had three positive effects: it has saved Walmart money, of course, but it also has reduced fossil fuel emissions and made both Conservation International and the company look smart and good. Today most Fortune 500 companies invest in offsetting their negative environmental impacts, and that is a step forward.

Yet tension remains in the tug-of-war between exploitation and preservation. Before departing Crossett, I joined GP forester Don Sisson to make a pilgrimage to the Morris Tree, a locally famous three-hundred-year-old Loblolly Pine that is 56 inches in diameter and 117 feet tall—famous because it is the largest and oldest remaining tree in the Crossett environs. The Morris Tree exemplifies what presumably many Arkansas pines looked like before the arrival of Columbus. Standing by the great tree, I was humbled by its grandeur, but also annoyed that commercial foresters had not demanded that their companies set aside representative plots of virgin forest within every large forest block that they harvested for timber. Such preservation plots, useful sources of seed stock and cuttings for future reforestation activities, would have been very beneficial to silviculture science over the long term and would have also had substantial educational value. For timber companies to set aside representative samples of virgin forest would have been another form of conservation offset, but, of course, it is now too late for that in Crossett or anywhere else in the

Southeast, where the virgin forests all have been converted to lumber, and those lands converted to monoculture plantations of young pines.

THE RCW AND THE ESA

On my last morning in Crossett, I head into the pinelands of Felsenthal National Wildlife Refuge. I hope to photograph the Red-cockaded Woodpecker, a species I have not seen since 1992, when I visited a pine plantation managed by International Paper in northern Florida. As I search for the bird, many hundreds of dragonflies patrol the gravel roads. So many odonates in one place! They remind me of the butterfly effusion I saw on Mill Road in the Tensas.

The Red-cockaded Woodpecker (or RCW, as forest managers call it) colonizes only old stands of living pines that have a relatively clear understory produced by periodic wildfire. Only a bit larger than a Downy Woodpecker, the RCW has a big white cheek patch and abundant black and white barring on its back. The adult male displays a small narrow red slash behind the eye. It's a small bird, but it has a big story.

The RCW is an important part of the southern pine ecosystem because family groups of the bird excavate large numbers of nesting holes across their home range, perhaps to give group members a range of night roosts that help them avoid predators. These drilled-out cavities are subsequently used by other woodpecker species, Brown-headed Nuthatches, Great Crested Flycatchers, Eastern Bluebirds, flying squirrels, and even tree frogs.

The RCW's global population is now a mere 1 percent of its estimated presettlement population. Historically a specialist inhabitant of Longleaf Pine savannas, this ecologically sensitive species declined drastically with the conversion of virtually all the 90 million acres of its fire-associated old-growth Longleaf Pine habitat in the eastern part of the country. Its classification as an endangered species by the U.S. Fish and Wildlife Service in the 1970s created a firestorm of controversy among the large corporations that managed millions of acres of pine plantations between North Carolina and Texas. Keeping the RCW from extinction (and satisfying the requirements of the Endangered Species Act, or ESA) has required millions of dollars of investment by state, federal, and corporate entities. Today many southern pine forests are young and an absence of fire has created dense pine-hardwood mixes, but the woodpecker requires trees older than eighty years and an open midstory without the successional hardwoods that sprout up in the absence of fire. Things are complicated still more by the species' social system, in which each breeding pair is assisted by offspring of a preceding year (called helpers); by each family group's territory size (exceeding a hundred acres); and by the fact that even with the assistance of helpers, RCW reproductive output is not high. The RCW is one of an array of threatened species in the United States that is *conservation dependent*, meaning that its survival depends upon substantial ongoing intervention by humans (we'll hear about another such conservation-dependent species—the Kirtland's Warbler—later in the book).

The Red-cockaded Woodpecker is a flagship species of the Endangered Species Act. Signed into law in 1973, the ESA mandates a strict set of rules for the management of lands holding species that the U.S. Fish and Wildlife Service identifies as endangered. For corporations that depend on exploitation of habitat-based resources, the presence of an ESA-listed species, such as a colony of RCWs, can have serious negative financial impacts. In the case of the timber and pulp enterprises, the presence of the woodpecker in a company's piney woods requires the development of a management plan to ensure the protection of the birds. As with the Spotted Owl in the Pacific Northwest and the Piping Plover along the beaches of the East Coast, local citizenry came to hate what they thought of as "job-killing" wild creatures fostered by the ESA. Communities disliked the federal mandate that required local action for endangered wildlife; it was expensive and cost jobs. But over time, businesses have adapted and recovery plans have been implemented, and today the Red-cockaded Woodpecker is protected on federal, state, and private lands throughout its range across the South, and the species is no longer in decline. Two remarkable technical innovations have aided the RCW's recovery from the brink. The first is wholesale translocation of family groups from areas with high densities of the birds to other areas of suitable habitat that lack the species. Birds are trapped at night, when they are in their roost holes, and safely released at the new site. Their transition to their new home is aided by the second innovation: the development of an artificial nest cavity, manufactured of molded fiberglass. These nest cavities can be readily placed in trees throughout a site that will receive a translocated colony to smooth their transition to a novel patch of pine woods. The RCW, of course, will never return to its original abundance, but with the mandated assistance provided by the ESA, it will continue to survive in its little colonies scattered through the piney woods of the South. It is a success story for the ESA.

Yet finding the RCW at Felsenthal wasn't as easy as I thought it might be. Instead I found an abundance of other woodpeckers, especially Red-headed and Pileated. Finally, stopping at a likely tree plot, I heard a telltale high-pitched sneeze. I looked up to see a single bird in female plumage busily scaling pine bark in search of food. She ignored me and my tripod. Periodically, she headed to her nest hole and fed offspring, which I could hear squeaking but could not see. They must have been quite young, as more mature nestlings typically poke their heads out of the nest hole to grab food from the parent's beak. The hole, high in an old Loblolly Pine, was made obvious by the big, messy swath of milky yellowish sap that covered the bark around its perimeter. The adult birds scar the trunk to produce these sap effusions in order to keep predatory snakes from entering the nest.

FELSENTHAL AND THE NATIONAL WILDLIFE REFUGE SYSTEM

A few miles west of downtown Crossett, Felsenthal is among the federal properties where RCWs are being protected and managed. Established in 1975, the sixty-five-thousand-acre refuge boasts an abundance of wetland resources, including Felsenthal Pool and sections of the Saline and Ouachita rivers. The lowest-lying sections of the refuge support seasonally flooded swamplands that give onto cypress and hardwood bottomlands. The uplands are dominated by pinelands. Historically, the Caddo people occupied the area, and important archaeological sites are well preserved within the refuge.

Felsenthal provides major local habitat for twenty species of wintering waterfowl, Blue-winged Teal, Black Ducks, Gadwalls, and Ring-necked Ducks among them. Bald Eagles concentrate at the refuge's wetlands in winter, and its upland forests serve as important breeding habitat for some Neotropical songbirds and as productive areas for passage migrants headed to more northerly breeding grounds. Its small population of Black Bears is one of only a few such populations remaining in the Deep South.

The National Wildlife Refuge System, of which Felsenthal is one unit, includes 521 refuges and more than 93 million acres of wildland habitat protected and managed for wildlife and game across our fifty states. Twenty million acres of this system have been declared as "wilderness" under the Wilderness Act of 1964. Far more so than the national parks, the national wildlife refuges are important wildlife habitat for migratory birds, especially for waterfowl and Neotropical songbirds.

The Mississippi Flyway is home to nearly a hundred of the refuges, encompassing nearly five million acres of protected wildlands. Most refuges here not only protect important wetlands resources, but also, as at Felsenthal, include an array of other productive wildland habitats useful for migratory birds. Most, but not all, of the refuges welcome visitors and provide trails and drivable wildlife loops so that birders, nature photographers, and the curious can get a look at America's natural patches. In season, fishing and hunting are permitted in designated sections. I would visit twenty national wildlife refuges—each a critical component of wild America—during my backroads journey.

MINGO AND JOINT VENTURES

My next destination is Mingo National Wildlife Refuge, a 21,600-acre reserve situated where the eastern edge of the Missouri Ozarks meets the northern extension of the Mississippi Delta—a perfect place for migratory

birds. I plan to meet American Bird Conservancy field scientist Larry
Heggemann here, early in the morning, a few days after I depart Crossett.

Mingo features all the goodies—cypress swamp, expansive marshy
wetlands, grasslands, oak bottoms, and hilly and rocky upland forest.
Centered on an ancient abandoned channel of the Mississippi River,
the reserve also has a twenty-five-mile wildlife loop, perhaps the longest in any national wildlife refuge.

Heggemann, with his thirty-plus years as a conservationist, is
an expert on Missouri wildlife and a perfect guide at Mingo. More
generally, I also wanted to learn about his work with joint ventures,
or JVs: regional institutional partnerships working to conserve
migratory songbird habitat in the various Bird Conservation Regions
across North America. These regions are ecologically distinct areas
with similar bird communities, habitats, and resource management
issues, and they were delineated by the North American Bird Conservation Initiative, a continent-wide partnership that includes state and
federal government agencies, nonprofit organizations, corporations,
and tribes. JVs use state-of-the-art science to ensure that a diversity
of habitats are available to sustain migratory bird populations. In
the United States, eighteen habitat-based JVs address bird habitat
conservation issues within their identified geographic zones. Four
habitat JVs focus on ecosystems in Canada. And three species-based
JVs, all with an international scope, further the scientific understanding needed to effectively manage a species or a group of species (the
Black Duck, Arctic geese, and sea ducks). JVs have a long history of
successfully leveraging public and private resources to draw partners
together to focus on regional conservation needs. Since the first JV
was established in 1987, JV partnerships have leveraged government-
appropriated funds to help conserve 24 million acres of critical habitat
for birds and other wildlife.

Heggemann, I learned, is responsible for promoting habitat management, land protection, and policies and programs beneficial for

birds of conservation concern in the Central Hardwoods Bird Conservation Region (CHJV), which includes parts of Arkansas, Missouri, Illinois, Kentucky, Tennessee, and Indiana. Heggemann works closely with state and federal agencies, NGOs, and other partners to seek opportunities to restore and manage natural communities that are critical to the needs of priority bird species on both public and private lands.

The particular species that are conservation priorities for the CHJV were identified in assessments performed by the North American Bird Conservation Initiative. Several hundred species of birds depend on habitat in the CHJV during critical periods of their life cycles. Many breed or overwinter here, while others stop over during migration between breeding and wintering grounds. Some species are doing well, but populations of others are exhibiting long-term declines. Species with the greatest need of conservation attention typically suffer some combination of vulnerabilities, such as a relatively small range, a small overall species population, or a reliance on a habitat under threat.

At 6 a.m., Heggemann and I were alone in the parking lot of the Mingo refuge visitor center until another car arrived and two birders popped out. Serendipity had brought us Mark Robbins and another expert birding colleague, who were doing a four-day bird survey of the state. I had first met Robbins, senior author of *The Birds of Missouri*, more than three decades earlier at an ornithological meeting in Philadelphia, and now Heggemann and I tagged along with these two top-gun birders. Robbins has a phenomenal ear and knowledge of songs, calls, and chip notes.

Birding alongside local experts of a certain age is special for several reasons. Of course, they know the birds of the area, but over the years, they have also visited all the region's nooks and crannies and divined the best spots for particular species, knowing when and where to look for each avian rarity. Moreover, they have lots of stories to tell. Heggemann and I knew we had stumbled upon an ornithological goldmine, and we mined this rich vein for all it was worth.

With help from Robbins, we recorded lots of thrushes and vireos, and more than a dozen migrant warblers to boot. I was able to list quite a few passage migrants at Mingo: Chestnut-sided, Magnolia, Bay-breasted, Tennessee, and Black-throated Green Warbler, plus Northern Waterthrush. These six were headed north, mainly to Ontario—my northernmost destination. As the birds and I moved north, I heard more and more species in song, which aided their discovery. But I was finding that the passage migrants, although present, were scattered thinly throughout the abundant habitat. It was like an Easter egg hunt: I needed to look under every bush and in every treetop to find the quarry. Teale's vaunted waves of migrants, of which I had dreamed, seemed no longer to exist.

Among the warblers I tallied with Robbins was a new quest bird: the Yellow Warbler. This specialist of pasture edges and willow swamps is a commonplace open-country warbler that rivals the Common Yellowthroat in continental abundance. The male is olive-backed and rich yellow elsewhere, with an abundance of rusty orange streaks on its breast. The Yellow breeds from northern Georgia to Alaska and Labrador, winters from the Yucatán to northern South America, and is a species most birders come to take for granted because it is a vocal breeder just about everywhere. It is a true rural roadside warbler, its bright song heard while one drives down country roads. Mingo, with its marshlands and openings, is prime breeding habitat for this species, but it is surprising that I hadn't recorded the bird earlier in my journey.

SONGBIRDS' INTERNAL GPS

When a Neotropical songbird migrant such as the Yellow Warbler passes over the Gulf and up through the southern United States to a breeding ground like Mingo, it must rely upon a sort of biological global positioning system. Migrant songbirds, as we noted earlier, not only have a "compass" that helps them distinguish north from south; they also possess a map sense that helps them navigate to a precise location on the earth's surface. With these two tools, the migrating bird can get where it wants to go.

The migrant bird's GPS system remains something of a mystery to scientists at this time. Yet we can understand some of its components, including its four central tools: magnetism, smell, low-frequency sound detection, and a bird's powerful memory of places and routes. Experimental evidence suggests that birds may use the earth's magnetism to detect their location on the globe, based on how the magnetic lines of force alter in declination based on distance from the equator. The closer one is to the North Pole, the more that magnetic lines of force trend toward horizontal. Nearer the equator, they are much declined due to the relative position of the North Pole and the spherical nature of the earth.

Moreover, studies of homing pigeons, European Starlings, and swifts support the remarkable notion that birds employ their sense of smell to detect location. This idea is not so far-fetched—recall that migrating salmon return to their natal stream by detecting the unique scent of its water, even when they're in the ocean. Some mammals' sense of smell is also keen: witness the dogs trained to locate hidden drugs or land mines. Such a skill could be very useful for adult birds returning to a breeding or wintering site.

Birds also can detect infrasound—very low-frequency sounds—which they use to locate known topographic features that produce distinct sound signatures (such as wind striking mountain ranges or waves striking coastlines). Animals' use of sound for navigation is probably more widespread than we know, mainly because humans lack this capacity. Think of the impact, for example, that the U.S. Navy's underwater sound propagation experiments have had upon populations of whales and other cetaceans.

Finally, older birds' ability to remember places and earth features may allow them to retrace routes year after year, in the same way that we remember the details of places we visited in decades past. It is clear that learning is a major part of the map sense, for young birds are unable to make their way accurately to a specific site, whereas adult birds can do it with uncanny precision. Many field experiments have translocated birds hundreds or thousands of miles from their nests.

The displaced birds were able to return with remarkable speed to their nests.

Experiments have proven birds' ability to take GPS-like actions, and we will speak more about these faculties when looking at research on thrush migration in a later chapter. In reality, however, there are still more questions than answers about birds' internal GPS. It's up to practicing research scientists to divine experimentally the finer details of the mechanism's construction and operation, and probably many amazing discoveries will be unearthed by future researchers working on an array of bird species across the globe. At this point, we must simply recognize the awesome navigational capacities of migrating birds.

RESTORING THE OAK GLADES

In the late afternoon, following Robbins's advice, I visited Cane Ridge, northwest of Poplar Bluff, Missouri, to look at an oak savanna restored by the U.S. Forest Service as part of local JV activities. In the protected landscape of southern Missouri, there is a heavy predominance of closed-canopy oak woods that shut out open-country birds. Here, west of the Black River, the managers of Mark Twain National Forest were opening up patches of oak forest to provide breeding habitat for American Woodcock, Red-headed Woodpecker, Blue-winged Warbler, Prairie Warbler, and Yellow-breasted Chat, among other open-country migratory birds.

This area was restored through selective clearing followed by managed fire to open the canopy and allow grassy understory to attract early successional bird species. Historically, this process happened naturally through the action of fire and large grazing ungulates (such as Bison and Elk), but these days, the management happens largely by mechanical means and follows a strict plan of intervention. At Cane Ridge, the plan had worked—I saw all the birds listed above. Such a prescription is needed in many more places.

While I walked the heavily altered habitat, an Olive-sided Fly-catcher sang out its *quick-three-beers*. This rare migrant species was headed north to some boreal bog in the North Woods—and it was the first Olive-sided Flycatcher I had seen on my road trip. It was a reminder both that I was soon to begin the northern half of my journey and that active habitat management can provide benefits for passage migrants as well as a wide range of breeders.

I added two new quest species at the Cane Ridge oak glades that I visited in Missouri: the Prairie and Blue-winged Warblers. The Prairie Warbler, which specializes in open, gladelike formations and old fields with scattered small trees, is a handsome bird with a breeding range almost entirely confined to the eastern United States. All yellow below and olive above, it is enlivened by black flank markings, a patterned face, and reddish streaks on its mantle. Its song, a rising series of buzzy musical notes, sounds like an energetic version of the song of a Field Sparrow, a species with which it can often be found. The Prairie is a partial migrant, with both breeding and wintering birds resident in Florida and the remainder wintering in the Caribbean.

The Blue-winged Warbler, a specialist of old field regrowth and shrubby clearings in disturbed woodlands, is a typical Neotropical migrant, wintering in the Caribbean and Central America and breeding in the eastern and central United States. Interestingly, the yellow and blue-gray Blue-winged hybridizes with its rare sister species, the Golden-winged Warbler, where the two species meet. The by-products of these crosses, informally called Brewster's and Lawrence's Warblers, are quite rare and create considerable excitement when they are encountered by birders. These two hybrids were originally described as novel species but now are understood to be the product of mixing of two closely related species.

Both Blue-winged and Prairie Warblers are uncommon because of their reliance on ever-changing early successional habitat. Breeding populations of the two species have to shift over time to track the movement of successional habitats across the landscape. Moreover, early successional habitats are becoming less common in the twenty-first century because of intensifying land-management practices.

CACHE RIVER

*From Missouri, I head east across the Mississippi to southern Illinois,
passing through the historic riverside city of Cape Girardeau (pronounced
"Jer-AR-doh"), Heggemann's home town. The bottom of Illinois, where the
Ohio, Cache, and Mississippi rivers converge, includes the northernmost
bottomland swamp forest that retains a Mississippi Delta accent. Here in
the last of the great southern swamp country, I plan to meet Mark Guetersloh, natural heritage biologist with the Illinois Department of Natural
Resources, who provides management guidance for the Cache River State
Natural Area.*

I was first introduced to this area by ecologist Scott Robinson in May
1993, when he and I, along with conservationist David Wilcove, did a
Warbler Big Day based out of Robinson's field station at Dutch Creek.
Traipsing high and low through many sectors of southern Illinois's
Shawnee National Forest in search of wood warblers, we finished with
thirty-two species, the highest single-day count of warblers I have
ever been party to. During that incredible day in this little-known,
wetland-rich world, I was amazed by how verdant it is, and how filled
with birds and snakes and other wildlife. This is the case, of course,
because southern Illinois is where the Ozarks meet the northernmost
reach of the delta. This rural landscape also features a long list of
conservation lands. I planned to spend three days in the area, reac-
quainting myself with the wonders of this "last of the South," where
Bald Cypress trees grow in swamplands rich with water-loving Cottonmouth snakes.

In the early afternoon, I met up with Guetersloh, a visionary
forester and ecologist who knows a great deal about the ecosystems
of southern Illinois and who works on restoring the hydrology of the
swamps and wetlands of the Cache River basin. Together we toured
Heron Pond and Big Cypress, two wetland areas featuring cypress.
During our travels through the Cache environs, Guetersloh told me
that there are serious challenges to protecting the health of the Cache

River ecosystem, including management of water levels, damming, and the presence of a canal cut to drain the Cache into the Ohio back in 1915. Different local interest groups—farmers, conservationists, and hunters—have disparate visions about water use, and the future of the Cache seems to be in the hands of the courts.

The natural places I saw here remain impressive to the visitor from outside, with their imposing stands of Bald Cypress and patches of old-growth bottomland hardwood. Walking the boardwalk at Heron Pond, Guetersloh showed me two Cottonmouths in the dark water and pointed out the lovely high, musical peeping of a Bird-voiced Treefrog—bright green–backed, with lichen-patterned gray and white on its sides. Late in the afternoon, he took me to see a state champion Cherrybark Oak with a diameter of more than seven feet that stands some 100 feet tall, with a spread of 113 feet. The Cache River wetlands are home to ten other state champion trees as well as the national champion Water Locust, and the lower Cache is home to trees more than a thousand years old. John James Audubon, passing through here in the winter of 1810, wrote about these forests with admiration:

> Though the trees were entirely stripped of their verdure, I could not help raising my eyes towards their tops, and admiring their grandeur. The large sycamores with white bark formed a lively contrast with the canes beneath them; and the thousands of parroquets [Carolina Parakeets, now extinct] that came to roost in their hollow trunks at night, were to me objects of interest and curiosity.

Guetersloh made clear that conservationists have to fight the good fight to conserve all the natural benefits that these places offer migratory birds and native plants and animals. It is not simply a matter of determining the right path forward for nature, but also one of ensuring that the best intervention is undertaken even in the face of opposing political forces.

WARBLERS OF PINELAND, OAK GLADE, AND CYPRESS SWAMP

My visits to the piney woods and cypress swamps had delivered close encounters with breeding wood warblers as well as their passage migrant counterparts. As I reviewed the warblers I'd seen in these lands before I departed for the North, I recalled that many of the tall pinelands I visited had rung out with the sweet slurs of the Yellow-throated Warbler, which I first saw on its breeding ground at Caddo Lake. One of the southern breeders, in some places it prefers pines and in other places prefers bottomland sycamores and cypress. The Yellow-throated is an unusual warbler because it is a *partial* migrant. In most warbler species, the bird's breeding habitat is distant from its wintering habitat. But in the northern parts of the Yellow-throated's breeding range, populations of the species entirely depart south in winter. In contrast, in parts of the coastal Deep South, some Yellow-throated populations breed and winter in the same site. In many pine stands, I'd also heard the soft trill of the Pine Warbler, the pinelands specialist that I first found at Felsenthal National Wildlife Refuge. Another partial migrant, it is a summer visitor to northern North America but a year-round resident in the southern pinelands. And I'd seen the partial migrant Prairie Warbler in the Missouri oak glades, along with the fully migratory Blue-winged Warbler. My warbler count was now at fifteen, thanks to the advice of Mark Robbins.

I am now about to leave the South and embark upon the second half of my journey. I'll trade my zig-zagging route through the South for a northward-trending route up the big river, bound for the Mississippi headwaters, the Canadian line, and the Great North Woods. There will be little lingering and much more movement. As with the songbird migrants themselves, the northern half of my journey will carry me northward faster and via a more direct route: straight up the main stem of the river.

SIX

From the Confluence to the Headwaters

Early to Late May 2015

April is promise. May is fulfillment. May is a time when everything is happening, when life rises to a peak. May is the birdsong month.

—EDWIN WAY TEALE, *A Walk through the Year*

*I travel to eastern Missouri and Trail of Tears State Park, which sits atop
a bluff on the western bank of the Mississippi River, fifteen miles north of
Cape Girardeau. Encompassing 3,415 acres of hilly upland oak woods with
an understory of Redbud and Sassafras, the park marks the spot where
bands of eastern Native Americans, uprooted by government mandate,
crossed the Mississippi on the way west to a reservation in Oklahoma in the
winter of 1838–39. This tranquil place memorializes a tragic story, which
I have come to learn, as well as to check in on spring migration at this
important patch of green along the river.*

As I walked to the visitor center, a Wood Thrush and a Kentucky Warbler sang from the woods just behind the building. Above the road to the campsite, three Mississippi Kites—the slim blue-gray raptors of riverine lowlands—circled in the clear blue sky. Another Kentucky Warbler sang down in the glen, a species that here outnumbered its vocal counterpart, the Carolina Wren. Woodland thrushes overran the park and foraged beside the narrow forest roads. A Chestnut-sided Warbler—a passage migrant—sang its cheerful song, and the camp hostess told me she had been hearing a Whip-poor-will calling most nights. I chose a ridgetop campsite that was woodsy but bug-free, graced with an oak canopy from which a Great Crested Flycatcher on territory sang out *wheep!* over and over in the evening light. Migrant songbirds—both breeders and passage migrants—were here in force.

First thing the next morning, I biked out to the high, rocky Boutin Overlook. In the river below, a pusher tug guided a long barge upstream amid considerable river traffic. East across the Mississippi were the expansive, forested hills of Illinois, including the towering summit of Bald Knob, with its giant cross. Aside from the cross, there was minimal sign of habitation, merely a vista of rolling green forest. Much of that green space was Trail of Tears State Forest—the next stop after my stay here at this state park.

OPPOSITE: Canada Warbler

Here on the bluff top, I was greeted by various passage migrants: Northern Parula, Tennessee and Chestnut-sided Warbler, Rose-breasted Grosbeak, and Scarlet Tanager. The Mississippi Kites continued their display flights preparatory to nesting, and several thrushes appeared: Swainson's and Gray-cheeked (passage migrants) and Wood Thrushes and Veeries (local breeders). I wondered what this spot had been like during the Trail of Tears exodus, almost two hundred years ago.

THE TRAIL OF TEARS

In 1830, President Andrew Jackson lobbied for passage of the Indian Removal Act, which called for the forced relocation of populations of Native Americans living east of the Mississippi River. American settlers were pressuring the federal government to remove Native Americans from the lands of the Southeast; many white pioneers filtering into Native American territory wanted the government to make these lands available for their own settlement. Although the effort was vehemently opposed by many, including Congressman Davy Crockett of Tennessee, Jackson was able to gain Congressional passage of the legislation, which authorized the government to extinguish Native title to lands in the southeastern United States.

The ensuing Trail of Tears exodus was a series of government-mandated relocations of remnants of various Native American nations from their ancestral eastern homelands to an area west of the Mississippi River that was designated as Native Territory (now Oklahoma). At the time of the forced migration, a few Native Americans living in the Southeast managed to remain on their ancestral homelands; for example, today some Choctaw still live in Mississippi, some Creek in Alabama and Florida, and some Cherokee in North Carolina; a small group of Seminole moved to the Everglades and were never uprooted by the U.S. military. But Jackson sent the vast majority of Native Americans west.

In 1831, the Choctaw became the first Nation to be dislodged. Their removal served as the cruel model for all future relocations.

After two wars, many Seminoles were removed in 1832. The Creek removal followed in 1834, the Chickasaw in 1837, and last the Cherokee in 1838. By 1839, forty-six thousand Native Americans from the southeastern states had been forced from their homelands, thereby opening twenty-five million acres for white settlement.

The term "Trail of Tears" originated from a description of the removal of the Cherokee Nation. While some Cherokee migrated voluntarily, more than sixteen thousand were forced out of their homeland against their will and made to march to their destinations by state and local militias. In the winter of 1838–39, a long procession of wagons, riders, and people on foot traveled eight hundred miles west to Tahlequah, Oklahoma. Most of the Cherokee made their way through Cape Girardeau County, now home to the state park. Floating ice stopped some of the attempted Mississippi River crossings, so the Cherokee bands had to set up camp on the riverbank. While waiting to cross, the Native Americans endured rain, snow, severe cold, hunger, and disease. Many women, children, and elderly people died; it is estimated that more than four thousand Cherokee lost their lives in this march of tribal decimation. Trail of Tears State Park, part of the Trail of Tears National Historic Trail, preserves the native woodlands much as they appeared to the Cherokee as they passed westward. These Native Americans had a special reverence for the animals that shared their land. My own travels encountering birds and mammals intersected at a number of places with those of Native Americans past and present, but how different my trek was from their tragic journey over the past two centuries.

OAK FOREST MANAGEMENT

A bit northeast of Trail of Tears State Park, and across the Mississippi, lies little-known Trail of Tears State Forest, in Illinois, which I reached after crossing the river at Cape Girardeau and wandering backroads beset with migrating turtles of various species. I set up camp on a sharp ridge cloaked in oak forest, much like the Ozark forests I had seen west of Mingo in Missouri. In fact, the Ozarks ecological region spans five states, including slivers of easternmost Oklahoma and southeastern-most Kansas, good chunks of northwestern Arkansas and southern Missouri, and a portion of southern Illinois.

Trail of Tears State Forest comprises more than five thousand acres of hilly upland forest, and I was probably the only person camped in it at this time, because it was midweek in spring. I saw Sugar Maples growing in the woods—a botanical signpost telling me I was easing into the northern half of my journey. Barred Owls hooted in the dark, and a long train rolled by in the distance at around 10 p.m. It must have taken twenty minutes for the string of cars to pass—the loud trundling on the rails and the periodic tooting of the locomotive's whistle brought on musings of a time in my childhood when railroads ruled and I heard the sound of trains every night.

In this state forest, I planned to learn about the challenges of mid-country forest management. Tracy Fidler, of Shawnee Resources and Development (a local nonprofit), had agreed to show me around, with guidance from Illinois state foresters David Allen and Ben Snyder. The trio are dedicated to the foresighted management of the forests of southern Illinois. It turns out that the mature oak-hickory forests that have long dominated the central hardwoods area of the country are reaching an ecological dead end, neither regenerating nor properly supporting an array of threatened migratory songbirds, due to the absence of periodic fire in the ecological regime and the resultant clos-ing of the forest canopy. Smokey Bear perhaps has been too successful in halting fire in America's forests.

Hunters and birders love oak-hickory forests, which attract both game birds and songbirds. The problem is, in the absence of natural regimes of fire and canopy disturbance, these mature oak-hickory woods will slowly but inexorably transition to less productive maple-beech woods. Expert forest managers such as Allen and Snyder man-age these forests to foster the healthy recruitment of new generations of oak and hickory to replace those in the canopy today. This forest needs active disturbance to bring about the succession of young oak and hickory seedlings into canopy trees. This seemed counterintuitive, but these eighty-year-old forests are not replacing themselves. That's why the foresters need to step in and take action.

Here is the true story of these oak-hickory forests. Recall that after the Civil War, southern Illinois probably was entirely deforested because of the chronic impacts of widespread agriculture, charcoal production, and the cutting of remnant woods for timber. Slowly, over a number of decades, farmers abandoned the unproductive hilly lands, which regenerated to old fields, then scrub, then woodlands, and finally forests. What stands here today is a direct result of this *single* historic cycle of succession from bare fields to mature forest. In earlier generations, recurrent fire events, small-scale agriculture by Native Americans, and other patch disturbances generated a mosaic of woodland and oak savannas that supported a wide array of habitats. Today, closed-canopy forest dominates, without the fire or patch dynamics that would keep it a diverse mosaic. Larry Heggemann and the Central Hardwoods Joint Venture seek to create a mosaic of woodland types, from the sunny and savannalike oak openings I'd seen at Cane Ridge in Missouri to the mature oak-hickory forests I saw here at Trail of Tears State Forest in Illinois.

As we toured a 925-acre demonstration area in the state forest and looked at a number of plots pre- and post-treatment, I gained understanding of what needs to happen here to generate fresh habitat for the Cerulean Warbler, Wild Turkey, and other local wildlife specialties. Proper management requires a combination of canopy thinning, midstory removal, and controlled burning. It is an expensive proposition, but it's necessary to restart the natural disturbance regime that molded these forests in centuries past. Of course, there has been a fair amount of pushback from the general nature-loving public, who often see the existing forest as "pristine" and "natural," although, of course, it is neither—it is the specific dead-end product of local human history, and it requires the human touch in order to provide the greatest benefit to biodiversity and birdlife. The public needs to learn that "disturbance" can be *good* as well as bad, depending on the scale and context.

Certainly, the devil is in the details. A field study by Aaron Gabbe and colleagues in southern Illinois has demonstrated that the threatened Cerulean Warbler preferentially forages in an uncommon species of tree: the Shellbark Hickory. It is a large-seeded bottom-land forest species that has difficulty recolonizing logged-over lands because its big seeds are not as easily dispersed as those of the many more common small-seeded species, such as the maples, oaks, and beech. Gabbe's research suggests that fully functioning forest ecosystems require the regeneration of Shellbark Hickory in order to provide ecological benefit to the Cerulean Warbler. The take-home point: in some cases, good management means more than simply setting aside land and keeping it undisturbed and free of fire.

After our field seminar, Fidler and I headed to Dixie Barbecue, a favorite local dining spot in Jonesboro, Illinois, where slow-smoked pork is sliced very thinly, grilled until slightly crisp on the edges, put on a warmed bun, and topped with a secret homemade BBQ sauce. That, with a cherry Coke, makes a fine downhome lunch. Aside from its barbecue, Jonesboro is famed as the site of the third Lincoln-Douglas debate, held on September 15, 1858. The seven "Great Debates" set Republican challenger Abraham Lincoln against Democratic incumbent Stephen Douglas in the race for a U.S. Senate seat in Illinois. Slavery (including the Missouri Compromise and the *Dred Scott* case) was the dominant topic in the debates. Unlike the soundbite format of current debates, the 1858 exchanges allowed one candidate to speak for sixty minutes, the second for ninety minutes, and then the first candidate was given a final thirty minutes to respond to the words of the second speaker. Lincoln lost that election to Douglas. Afterward, with free time on his hands, Lincoln collected and published the transcripts of the debates as a book, which was very popular and assisted with his election as president two years later.

URBAN BIRDING

My next field activity is scheduled for downtown Saint Louis, an urban birding hotspot. I have plans to birdwatch with local naturalists in one of the popular downtown parks of the city. Although I'm including city birding

*in this largely rural journey, I do not plan to stay in Saint Louis. Instead
I'll camp in Pere Marquette State Park, north of the city in Grafton,
Illinois.*

Driving north on Interstate 55 toward the city, I encountered the first
road-killed Coyote of the trip, plus a couple of road-killed Armadillos.
Both species have been on the move in the East in recent decades. The
Coyote has colonized much of the suburban East Coast, even appearing in city parks and preying upon local residents' domestic pets.
The Armadillo, confined to Mexico in the 1880s, continues its march
northward into America's heartland, but not without abundant road
mortality. On a happier note, a Pileated Woodpecker, high in the blue
sky, crossed I-55, the big black bird flashing its white underwings as it
undulated gracefully from one woodland patch to another.

On my way to Pere Marquette State Park, I passed through urban
Saint Louis, which the summer before had been rocked by the Fer-
guson, Missouri, riots precipitated by the police shooting of Michael
Brown. I made my way through downtrodden northern sections of
Saint Louis toward the bridges crossing the Missouri and the Missis-
sippi to Alton, Illinois. Passing the high, pale-gray limestone bluff on
the Illinois side of the Mississippi, I followed the Great River Road
twenty-one miles upstream to Pere Marquette State Park, where the
Illinois River joins the Mississippi. It then became clear to me that
Saint Louis is in this spot precisely because the three great rivers—the
Missouri, Mississippi, and Illinois—come together here, smack dab in
the middle of the country.

Pere Marquette State Park, at eight thousand acres the largest in
the state, has access to the Illinois River and a large boat basin. The
park is lent a rustic ambience by a number of 1930s-era stone-and-log
buildings built by the Civilian Conservation Corps, including a large
guest lodge facing a set of pretty stone cabins. The tenting ground, a
bit uphill, had trees, but overall this seemed a rather suburban park,
influenced by its proximity to the big city just to the south.

The next morning, May 14, I rose at 4 a.m. and drove downtown to meet my birding party: knowledgeable naturalists Christian Hagenlocher, Brad Warrick, Jacob Warrick, and Garrett Sheets, local residents who had answered my emailed call for birding guidance in the city. Together we headed for Tower Grove Park, in the middle of Saint Louis, and spent two hours birding for migrant songbirds. It was one of those cool and wet spring days that birders learn to appreciate—gloomy weather quite often produces excellent urban spring migrant birding. Small flocks of White-throated Sparrows foraged under every patch of shrubbery, their presence telling me that I was still near the front edge of the northbound wave of songbird migrants.

Birding downtown parks is best during the spring migration, because cities' vast expanses of concrete and asphalt make every small patch of green vegetation attractive to migrants at the end of a long night of flying. New York's Central Park and Chicago's Magic Hedge stand out as the most famous, but in fact most eastern and midwestern cities hold green spaces that bring in migrant birds, as well as birders, in numbers. Today the birding was good: our little team recorded fifty-four species, including a Black-throated Blue Warbler (a rare passage migrant this far west), vocalizing Olive-sided and Yellow-bellied Flycatchers, and a total of thirteen species of wood warblers passing through town.

Later in the morning, I set the GPS for West Alton, a noted birding destination on the southwest bank of the Mississippi. Waiting at a downtown stoplight, I looked down an alley to see several feral cats hanging around what looked like a feeding station: a site periodically provisioned by some kindhearted person. Cat lovers maintain hundreds or thousands of feral cat colonies in cities and towns across the United States, an act of kindness with unintended consequences. Outdoor cats kill more than two billion birds each year in North America and can carry and transmit serious diseases, such as toxoplasmosis, to humans. Feeding feral cats does not halt their hunting of birds and mammals. Cats are efficient natural predators, and their introduction to novel landscapes has contributed to the extinction of thirty-three species of birds around the world. And for every wild bird killed, cats kill two to three wild mammals (chipmunks, rabbits, voles, shrews). The harm to vertebrate wildlife populations is substantial. Studies have also shown that feral cats live unhealthy, brutish lives, with none of the pleasures known to indoor cats properly cared for by their owners. It turns out that the trap, neuter, release, and feed movement is really not humane treatment for these creatures, which lead mean lives and which are destructive in urban and rural landscapes. The American Bird Conservancy, through its innovative Cats Indoors! Program, has worked for more than a decade to educate cat owners and cat lovers about the proper stewardship of their pets and to foster state and national policies that protect wildlife from the depredations of wild-roaming cats.

ABC works on other urban/suburban bird issues besides cats, two of which are particularly important for migrant songbirds. The nighttime illumination of tall urban buildings leads to the maiming or death of many spring migrants, especially on foggy and rainy nights, as do lighted transmission towers. The threat is exemplified by a kill of more than four hundred songbirds of twenty species, including many migrant wood warblers, at a lighted building in Galveston, Texas, in early May 2017. ABC is working with cities to alter building and tower lighting to reduce bird deaths. Reflective windows also kill many birds, especially in suburban habitats where windows face a mix of lawn and woody vegetation, and so ABC has worked with companies to create specialized window tape that discourages bird collisions. Cats, windows, and lighted buildings are just three of the threats that migratory birds face when passing through urban and suburban landscapes.

The confluence of the Mississippi and Missouri rivers, on the northern verge of Saint Louis in West Alton, includes verdant wetlands and bottomlands, even though it is adjacent to a sprawling urban center. Because of the area's flood-prone nature, the Army Corps of Engineers has created all sorts of impoundments and flood-control structures that, happily, are attractive to birds. After a brief visit to the Audubon Center at Riverlands, just north of town (highlight: Blue Grosbeak), I wandered about on what is essentially a peninsula between the two big rivers and ended up, without knowing where I was going, at Edward and Pat Jones Confluence State Park, where the Missouri and Mississippi meet. I walked out to the wooded point where these two streams of silt-laden water come together. It is awe-inspiring to stand at the narrow point of land where these great rivers collide, and where geography and history and nature come together.

Birding on the wooded point yielded a Gray-cheeked Thrush, an Indigo Bunting, a White-crowned Sparrow, and both Baltimore and Orchard Orioles. Several vocal Warbling Vireos and Northern Cardinals hung out in the small triangle of Cottonwoods that forms the heart of the park. Out low over the muddy and turbulent Missouri, Chimney Swifts and Purple Martins hawked insects. It was gray and cool, damp and river-girt, and now I felt I was truly launching into the northern sector of my journey. Goodbye, southlands.

At 5 p.m., back at Pere Marquette State Park, the rains had finally finished. The sun began to shine, and the black flies to swarm. They know exactly where to land and bite to generate the greatest effect. I slathered bug repellent on my neck and temples and behind my ears—the target zones of these devilish little dipterans. I looked through my food supplies and discovered that I had lost yet another loaf of bread to marauding Raccoons. When I had chased off last night's thief, he'd also been sampling my tortilla chips.

In the early evening, before cooking dinner, I tallied a Black-capped Chickadee here at the campground, a species confined to the northern half of the country. Tower Grove Park, downtown, hosts the southlands-dwelling Carolina Chickadee as well as some hybrids between the two species, but at Pere Marquette, it is all Black-capped. The local bird fauna was signaling my arrival in the North. I had now completed a bit more than six weeks out on the road—almost half the journey's allotted time.

The thrushes that I had seen at Edward and Pat Jones Confluence State Park, and spotted in numbers elsewhere in the midcountry woodlands, are ideal subjects of field study because they are large enough to carry tiny radio transmitters on their backs, which allow researchers to follow the migratory movements of individual birds. This is a great way to learn about how songbirds navigate to their nesting territory in the North Woods.

William Cochran and Martin Wikelski have conducted three decades of radio-transmitter research, and their studies reveal the basics of thrush navigation. After each night flight, northward-migrating thrushes stop over in woodlands and feed until their fat levels are restored to preflight levels (which requires several days). In spring, these birds have a remarkable ability to put on the fat reserves needed to power migration as well as to handle the future demands of establishing a breeding territory. Stopover birds forage in an area only about a hundred yards in diameter. At the end of the day, before departing on their next flight, the thrushes calibrate their magnetic compass based upon where the sun sets or upon the plane of the polarized sky light that they detect overhead.

The birds migrate only at night, and only on nights when the air temperature is not too chilly and when the wind speed at ground level is less than six miles per hour. In flight, the birds beat their wings about six hundred times per minute. Their heart beats at around the same rate. They typically fly about thirty-five miles per hour (ground speed) but usually benefit from a tail wind. They fly until they deplete their fat reserves or until daybreak, whichever comes first, and then they drop down into another wooded patch with access to water.

Night flights last as long as eight hours, and the birds travel as much as five hundred miles each night but typically only half that distance. Perhaps most remarkable is that the thrushes keep a constant magnetic heading during their entire migration up the Mississippi. Once the northward-traveling birds reach the latitude of their breeding habitat, they switch to eastward- or westward-trending flights to locate their preceding year's nesting territory. The mainland migration (from the Louisiana coast to breeding habitat in Canada) takes on average about forty days, which includes about eight night flights and many additional days and nights of rest and foraging. The flight from Panama, where they winter, to Ontario entails 3.2 million wingbeats.

Unexpected findings include the discovery that overland migration imposes only a moderate energetic demand on the thrushes. The researchers also found that birds typically stop migrating and drop into a woods upon encountering a cold front. Most remarkable of all is that migrating thrushes change their orientation to fly *toward* a thunderstorm when lightning is visible. The most likely explanation for this phenomenon is that the thrushes want to stop over in a site that has plenty of water available (which is likely if a thunderstorm had recently passed). The most important result of recent studies of thrush migration is that the birds can successfully navigate across the continent by using a set of simple migratory decision rules, rendering what at first appears impossibly complex into the realm of the understandable. Some mysteries remain, of course. For instance, once the songbird arrives at the proper latitude, how does it determine whether its natal breeding site is west or east of that point? Which GPS-like cue provides that information? That riddle remains for researchers to answer.

EFFIGY MOUNDS AND THE BIRDS OF THE DRIFTLESS AREA

I next travel north to the confluence of the Wisconsin River with the Mississippi, site of Wyalusing State Park, where I will camp for several days. Along the way I stop at the restored prairie at Wapello Land and Water Reserve in Hanover, Illinois. In the early nineteenth century, this patch of prairie was a Native American village headed by Chief Wapello of the Fox (or Meskwaki) people, and the archaeological site reveals two periods of occupation, the oldest tracing to 1050 CE. North of Hanover, a Red-bellied Woodpecker flies across the road in front of the car, reminding me that I am at the northern edge of the range of this nonmigratory southern species, which has expanded northward over the past half century.

As I approached Galena, Illinois, in the northwestern corner of the state, I turned onto the Ulysses S. Grant Memorial Highway, which leads to the former president's hometown through verdant rolling country, a mix of woods and well-tended agriculture, and descends the high sandstone plain to Galena. Looking down, I saw what seemed like a New England college town, its church steeples rising out of maples. The best historic aspects of old Galena have been preserved and restored—it is one of those small and picturesque rural towns whose economy has adapted to attract tourists who love pleasing scenery, restored old buildings, tranquility, and good dining.

In midmorning, I arrived at the Wisconsin state line. The Mississippi has cut a deep valley here, and adjacent to the river the land is wooded and hilly; back up on the ancient elevated plain, away from the river, it is flat farmland. Here the farms were big and prosperous, each farmhouse surrounded by four or five major outbuildings. The neatness of the farms was impressive, but they were too tidy for my taste, one reason that on my way north I kept mainly to small roads nearer the river, which were greener and less tame. A lot of Black Cherry trees were in bloom, and the leaves of the commonplace Black Walnut trees were just starting to unfold; many trees on the exposed ridgetops had not yet fully leafed out. Spring was only just arriving here, although it was already May 18. The radio announced an expected predawn low of 39°F for the 19th, and the northwest wind was starting to blow, which would slow the northward songbird migration.

Wyalusing State Park perches atop a high bluff looking down on the deep valleys of the Wisconsin and Mississippi rivers, whose confluence lies just west of the park. I settled into a campsite set a bit back from the bluff, protected from the winds striking the northern face of the escarpment. As I erected my tent, I was greeted by the songs of Rose-breasted Grosbeak, Wood Thrush, and Tennessee Warbler. A female American Redstart gamboled about the campsite no more than a few feet off the ground, and a male Cerulean Warbler sang repeatedly in the forest canopy just above the picnic table. The birdsong was welcoming, but the approaching chilly weather was not.

Cerulean Warbler

In the afternoon, I visited Effigy Mounds National Monument in Harpers Ferry, Iowa, just across the Mississippi and upstream from Prairie du Chien, Wisconsin, the nearest town to the state park. Here, about a thousand years ago, Native American communities fashioned giant animal-shaped earthen mounds on the high plateau overlooking the river. Of its 206 surviving earth constructions from the mound-building culture, Effigy Mounds National Monument exhibits a range of examples, and I visited several of the most famous: Big Bear, Little Bear, and a line of simple conical mounds that resembles a necklace of beads, just back from Eagle Rock lookout. The stiff climb from the visitor center up to the main concentration of mounds was lightened by vocal songbirds in the woods, including numerous locally breeding Rose-breasted Grosbeaks and American Redstarts. Mature forest has grown up to engulf the mounds, which are scattered far and wide. The Park Service has cleared the mounds of woody vegetation and trees to make it possible to see the details of the ancient earth sculptures. One wonders what the environment looked like when these mounds were in active use. Were they set in woodlands or open country?

The Mississippian Mound-Builder culture spread out through the central reaches of this great watershed between 1100 and 1350 CE. The mound-building people here in eastern Iowa fashioned hundreds of mounds, some in geometric shapes and others that graphically represent turtles, bison, lizards, and, most commonly, bears and birds. Some tribal narratives held that the bear was the guardian of the earth and the bird the guardian of the sky. Excavation has shown that fire, too, was associated with many mounds.

Perhaps the most remarkable construction is the Marching Bear Group, a linear assemblage of ten bears plus three birds. The true meaning of these thousand-year-old mounds remains a mystery. They were constructed by people of the late Woodland period, hunter-gatherers who prospered on nature's bounty in the watershed of the Mississippi and mainly inhabited the environs of southern Wisconsin and Iowa. The last of the effigy mounds were constructed 850 years ago, when the people of the Oneonta culture came to dominate the region with large permanent settlements and new forms of pottery. Surveys of northeastern Iowa in the nineteenth and early twentieth centuries documented more than ten thousand prehistoric mounds in that state alone. By the mid-twentieth century, fewer than a tenth of these had survived the widespread development of the landscape. Effigy Mounds National Monument was established in 1949 to pre-serve a small remnant of this important archaeological legacy.

By 8 p.m., back at Wyalusing State Park, it was windy, cloudy, and cold. I had a hot meal and bundled myself in several layers of fleece, a woolen watch cap, and woolen gloves. High up here, bundled against the weather and looking down on the Mississippi, I was reminded of November days I'd spent at my favorite autumn hawk watch, north-west of Carlisle, Pennsylvania. The Wyalusing bluff top was as exposed as the typical ridgetop hawk watch, and just as chilly, too. Luckily, I had brought plenty of clothing in anticipation of cold in the far north. At my little campsite, I was serenaded by the singing male Cerulean

Warbler and several Rose-breasted Grosbeaks—signaling that in spite of the cold weather, today's movement of the sun across the sky had told the birds that yes, it was indeed spring.

The predawn chorus started around 5 a.m. as a bunch of American Robins gave a weird short song in the full dark and then shifted into their typical longer, warbling dawn song. Finally, other bird species chimed in as the cloudy morning broke. I emerged from my tent at 6 a.m., and the same male Cerulean Warbler sang over my shoulder as I ate breakfast in the cold and windy morning. The Cerulean was yet another of my quest birds on its breeding territory.

The Cerulean Warbler, a species in decline, is one of those migrants that everybody seems to search for. It is fine-looking and it dwells in the canopy of tall forest, which typically makes locating a Cerulean a challenge. The male, while it cannot compete with the beauty of the male Goldenwing or Blackburnian, is a demurely handsome creature, with sky-blue upper parts, white wing-bars, white underparts broken by flank streaking, and a neat, dark throat collar. The species breeds from northern Arkansas to southern Ontario but is found mainly in the heartland states of Wisconsin, Illinois, and Indiana. It prospers these days only in mature deciduous forest atop ridges, in places where oak and hickory trees grow tall (e.g., a place just like Wyalusing). To find one, a birder must listen carefully for the bird's evocative, rolling, and musical song, which gives me goosebumps when I hear it each spring: *tzeed tzeed tzeed tzeed ti ti ti zeee?*

I continued my hunt for novel migrants and local breeding songbirds. On this morning, I visited the restored prairie along the Wyalusing entrance road, highlighted by a number of educational signs discussing the restoration. Here I found five species: Ring-necked Pheasant, Red-winged Blackbird, Field Sparrow, Common Yellowthroat, and American Goldfinch. A cuckoo gave a cadenced series of *coo* notes in sets of three—it was a Black-billed Cuckoo, rather than the more common Yellow-billed. Hearing this uncommon and declining Neotropical migrant got my blood pumping. What other oddities might be around?

Standing in the restored prairie, I saw old pasture land in the distance, behind a little astronomical observatory set back from the entrance road. I wandered in to look for interesting grassland and edge habitats. Before going far, I heard the song of a bird I have heard only once before, in northern Texas. Here at Wyalusing, it was giving its rapid and buzzy zig-zagging song from a row of trees between two old fields. Its often-repeated song helped me locate the bird in a jiffy: a Bell's Vireo.

Absent from the East but widespread in the Great Plains, Bell's Vireo also occurs in the arid Southwest. Its plumage is plain, like a cross between a Warbling Vireo and a Philadelphia Vireo, and this species, like the Warbling, is distinguished by that remarkable song. The rapid-fire series of musical notes is unlike that of any other North American bird, and its habit of repeating its song over and over makes the bird all the more distinctive. This singing male was being harassed by a male Ruby-throated Hummingbird, which repeatedly dive-bombed the vireo. The vireo remained in its low tree, singing, while I photographed it at close range. I had hunted for the species at Mingo in Missouri, with guidance from Mark Robbins, but without success; what a surprise to find it here in Wisconsin, at the very northeastern fringe of its breeding range.

Sated with the sights and sounds of the vireo, I wandered farther from the main road into a series of large, unkempt, hilly fields surrounded by woods. Before long, I heard a single Henslow's Sparrow singing deep in the grass of one weedy hill. But it was midmorning, and this species is an early bird. I would have more luck seeing this elusive species if I returned very early the next morning.

I completed my midmorning birding tour of Wyalusing on bicycle. The state park supports oak uplands, prairie, and steep, forested bluffs that drop down to the two rivers, where wetlands and bottomland forest predominate. The tree community, from the top of the ridge to the bottomland, includes Black Oak, Shagbark Hickory, Black Cherry, White Oak, Red Oak, White Ash, Sugar Maple, Basswood, Ironwood, American Elm, Silver Maple, Eastern Cottonwood, River

Birch, and Black Willow. At the end of my bike ride, I tallied my morning count of warblers: ten American Redstart, seven Tennessee, five Common Yellowthroat, five Blackpoll, five Cerulean, two Kentucky, and a Chestnut-sided—five species of breeding residents and two passage migrants, but only the Cerulean an addition to the quest list.

At midday I reviewed the map and brochure for Wyalusing State Park and discovered, to my great surprise, that the park itself had held 107 Native American mounds when it was first surveyed in 1894. I wandered the roads and located a number of signboarded mounds, one in the form of a bear. The day fined up and the sun came out, but it remained windy and cool. Even at 2 p.m., I needed to wear gloves to work on my diary—the temperature was in the 50s. At the end of the day, I bathed in a shower house whose ambient temperature was 45°F. At 7 p.m., I'd finished dinner at my campsite, but that resident Cerulean Warbler male was singing its buzzy song every eight seconds nearby. Lots of other birds, too, still sang in the gloaming—orioles, wood pewees, and more.

Wyalusing and Effigy Mounds lie in the heart of the Driftless Area (also known as the Paleozoic Plateau), a region noted for its deeply incised river valleys. Primarily located in southwestern Wisconsin, the Driftless Area also encompasses portions of southeastern Minnesota, northeastern Iowa, and northwestern Illinois—an area of sixteen thousand square miles. The region's unusual terrain is the result of its having escaped the most recent period of North American glaciation. Retreating glaciers leave behind *drift*—silt, clay, sand, gravel, and erratics (loose boulders), plus unsorted material called till and outwash deposits from glacial meltwater streams. This area lacks all of the distinctive calling cards of recent glaciation; the only drift found here dates from more ancient glacial periods, reaching back five hundred thousand years. The Driftless Area is an eroded plateau with bedrock overlain by varying thicknesses of loess, or windblown sediments. And most characteristically, the rivers flow through deeply dissected valleys, along which the river bluffs stand tall and prominent. The sedimentary rocks of the valley walls date to the Paleozoic Era. The area has not undergone much tectonic action, and its visible layers of sedimentary rock remain horizontal. Considering how far south the last glaciers advanced, this glacier-free zone is a geographic mystery.

At 5:30 the next morning, I was back at the weedy fields behind the observatory, listening for Henslow's Sparrows. Within a few minutes, I heard the voices of four singing males in the rolling overgrown grassland, but I did not see a single bird. Henslow's is among many birders' holy grails, a will-o'-the-wisp that seems always just out of reach. I'd hunted for it unsuccessfully for several decades and only recently had gotten to know the bird first-hand. Most prevalent in the prairie lands of the upper Mississippi drainage—Ohio, Indiana, Illinois, Michigan, and Wisconsin—the bird is an old-field specialist that needs just the proper amount of thick, shrubby growth in a field to make it desirable habitat. Breeding populations of the species come and go as fields evolve through successional stages. Well-kept hayfields tended by farmers never have Henslow's.

Another reason for the difficulty of finding Henslow's Sparrow is its voice. Its quiet, slurred single note of a "song"—*sllinnk*—is easily overwhelmed by the cacophony of a morning chorus, when virtually every other bird species has a louder and farther-carrying vocalization. The only similar vocalists in the region are two other uncommon grassland specialists: the Grasshopper Sparrow and Le Conte's Sparrow. Yet another reason for the difficulty of finding Henslow's is its behavior. It sings mainly before dawn and after dusk, and it hides quietly deep in the grass for most of the day. Late-rising birders will not find many Henslow's Sparrows. This morning, I was here early, and I scored. But I had no luck with photography, given the poor light and the reclusive nature of the singing males. As a consolation, I did hear, once again, the noisy song of that male Bell's Vireo on territory back in the tree row next to the observatory.

From atop the grassy hilltop I also heard the *tatatatat—tatat—tatat* drum of a Yellow-bellied Sapsucker from a nearby woods, the distinctive staccato spring sound reminding me once again of the North Woods. The arrival of spring here could be seen in the abundant bloom of Black Cherry trees and Shadbush, as well as Northern Bush-Honeysuckle. Yet it remained cold: even at 8:30 a.m. it was 44°F, although not as windy as the preceding day. It occurred to me that I'd heard no frogs calling in the park—perhaps it simply was too cold for them.

As I brought my bicycle back to the paved main park entrance road, a Meadow Jumping-Mouse, with its very long tail, crossed the road in several prodigious jumps. I counted up the numbers of warblers I'd seen today: four Blackpoll Warblers and a single Myrtle Warbler. These two passage migrant species typically represent the earlier half (Myrtle) and the later half (Blackpoll) of the warbler migration season. Their presence here confirmed that I was right in the heart of "warbler spring" as it flowed ever northward. Later in the day I'd see several White Pine sentinel trees outside the park: still more harbingers of the Northlands.

CREX MEADOWS

As I travel north from Wyalusing, the GPS takes me to both sides of the Mississippi, from Wisconsin to Minnesota and back to Wisconsin. The sky is deep blue, the air clear and fresh. The spring sun begins to warm the air; this Friday, near the end of May, is one of those days we hope will last forever. I stop for breakfast in the little Minnesota college town of Winona, named for the legendary first daughter of Chief Wapasha of the Mdewakanton band of the eastern Sioux. Five colleges and universities are here: Winona State University, Winona State College, St. Mary's University, Minnesota State College–Southeast Technical, and the College of St. Teresa. I judge from this quick visit that Winona would be a great place to retire and perhaps teach a course every other semester. But I can't stay to explore the town's charms. I must move on, as I am on my way to northern Wisconsin to camp near Crex Meadows, a famous birding reserve in the northwestern part of the state, and well into wolf country.

I stop in Saint Croix Falls, Wisconsin, on the east bank of the Saint Croix River, in search of lunch. Loggers Bar and Grill fits the bill. I sample

local favorites: batter-fried cheese curds washed down with a Leinenkugel
beer brewed in Chippewa Falls.

In Grantsburg, Wisconsin, I headed to the visitor center for Crex
Meadows State Wildlife Area. Adrian Azar, a young birder from North
Dakota whom I'd met at Wyalusing State Park, suggested I see Crex
Meadows, singing the praises of the large protected wetland and
telling me that he himself would visit Crex Meadows on his way home.
That was recommendation enough for me.

The rangers at the center resolved my concern about finding an
available campsite over the Memorial Day weekend. They pointed
me to Governor Knowles State Forest, right on the banks of the Saint
Croix River, a National Scenic Riverway, which had an abundance of
open campsites. Here I'd spend the rest of the weekend camped along
the river and reveling in Crex Meadows' natural wonders, heightened
by the prevailing rain-free weather, unlike the typical cold, misty, and
rainy spring days of the North Woods.

Before setting up camp, I spent several morning hours checking
out Crex Meadows, some thirty thousand acres managed for wildlife
and nature. The area is part of the northwest Wisconsin pine barrens,
a large sand plain left from the retreat of the last glacier, eleven thousand years before the present day, and its extensive wetlands are
a remnant of glacial Lake Grantsburg (we are clearly out of the Driftless Area). When Native Americans lived here, the area was a fire-
prone brush prairie. The wetlands were drained in the 1890s, and in
the early twentieth century, Crex Carpet Company owned much of the
land and manufactured grass rugs here. The wildlife area was established in 1946.

Autumn migration here is famous for its Bald Eagles, Sandhill
Cranes, and many species of waterfowl. During the spring, birders
come from far and wide to witness the communal mating display of
the Sharp-tailed Grouse and to search for various rare marsh-breeding
birds. Current management focuses on restoring the wetlands and

From the Confluence to the Headwaters

prairies through deployment of a dike system, mechanical opening of closed forest tracts, and prescribed burns. Here is a prime example of the successful restoration of a major wetland resource after it had been essentially destroyed.

At my campsite above the Saint Croix River, I was greeted by the voices of an array of songbirds on their breeding habitat: Wood Thrush, Veery, American Redstart, and more. The campground was mainly young deciduous forest with a scattering of White Pines, and the tent site was bracketed by two big stands of Interrupted Fern, a plant I knew well from summers in the Adirondacks. In the late afternoon, as I organized my camp before making dinner, the sun remained high, a pleasant breeze blew, and I looked forward to spending all of Saturday morning exploring Crex Meadows. Abundant black flies swarmed but did not bite; my head net was not yet required.

I heard another song at the campsite, too: that of the Ovenbird, one of my quest breeders. It was surprising that I had not already located the species; this common and vocal bird breeds as far south as Arkansas, and winters from Florida to northernmost South America. Like the waterthrushes, it is a ground dweller, and it has the look of a sparrow: olive above and white below, with black streaks on the breast. An orange stripe adorns the top of its head, bracketed by two black racing stripes, and it sports a prominent white eye-ring. A common species of deciduous forests in spring, the Ovenbird makes the woods ring with its loud song: *teacher teacher teacher TEACHER TEACHER!*

Another voice of a different quest species came to me here as well: that of the Chestnut-sided Warbler. It is a gaily colored inhabitant of brushlands and the edges of northern swamps and cutover woodland openings, breeding north to central Canada and wintering south to Colombia. The male is mainly dark above and white below, with a distinctive yellow cap, a black mustache, and a chestnut stripe down the flank. Its cheerful and rapid chattering song is reminiscent of that of the Yellow Warbler.

At 5 a.m. Saturday, I was awoken by a loud dawn chorus of American Robins in the chill air (it was only 39°F). A croaking Common

Raven greeted me as I drove into Crex Meadows—the big black bird sailed right down the middle of the road and passed low over the car, its long, wedge-shaped tail prominent. All morning I traveled the huge reserve's extensive system of gravel roads, stopping to bird favorable spots. Standing on a dike cloaked in early-morning mist, I listened to abundant birdsong. Cranes were bugling. Canada Geese were honking. Sedge Wrens and Grasshopper Sparrows chorused lustily from the grass. A Sora gave its strange musical upslur from the marsh. An Alder Flycatcher sang *fee-BEE-oh*. Loons and Ring-necked Ducks moved silently over the glassy water.

And here I could see *daytime* migration! Dozens of small flocks of migrating Blue Jays passed overhead in waves all morning long. Their relentless flight northward reminded me of the migratory imperative—it's like the tide, not to be denied. A single Veery, another northbound migrant, crossed a large swath of open marsh. I hiked the Phantom Lake Trail, where the many oaks were just pushing out their tiny, pale leaves, whose translucency reminded me of the wings of a butterfly when it emerges from its chrysalis: the start of a new life in early spring, here in the north country.

Like a rapidly beating heart, the courtship drumming of a male Ruffed Grouse atop a fallen hollow log resonated from afar, so low-pitched that I felt it more than heard it. The characteristic cadenced North Woods drum of the Yellow-bellied Sapsucker joined it. And the wood warblers were in voice, too: Golden-winged, Nashville, Black-and-white, Tennessee, Yellow, Wilson's, Chestnut-sided, Common Yellowthroat, and Ovenbird. The woods trembled with birdsong. In the marshlands I heard the weird *pump-er-lunk* of the American Bittern—an uncommon marsh bird that is rarely seen except when in flight—and the spring courtship whinny of a Pied-billed Grebe.

Yet I failed to detect three birds that I had specifically hoped to see at Crex Meadows: it is famous for Yellow Rail, Le Conte's Sparrow, and Nelson's Sparrow, all of which breed in low, wet, grassy, freshwater marshlands near the U.S.-Canada border. All three are elusive, and they stayed true to form on this day. I saw plenty of Sedge Wrens and Grasshopper Sparrows and even a Clay-colored Sparrow, but not a single hint of the more elusive trio—the early bird does not *always* catch the worm.

Of course, I was eager to see other types of wildlife. The ranger at the visitor center had mentioned that a pack of wolves lurked near the northwestern corner of the reserve. I managed to locate wolf footprints in the sand but had no sightings of the animals themselves. I would have another chance for this wilderness wraith in Ontario.

By midafternoon, I was back at my campsite to bird in the cool, sunny weather. The campground was favored by species of early successional vegetation: I found Interrupted Fern, White Trillium, and flowering Shadbush, and I was pleasantly surprised to find no Poison Ivy. Male Chestnut-sided Warblers chased about the edges of the campsite, singing frequently. A couple of Eastern Cottonwood trees shed their fluff in bits that filled the air throughout the afternoon. A migrant Philadelphia Vireo sang from various of the taller aspens in the campground.

A vocal male Mourning Warbler moved about the tangled thicket in the adjacent campsite, and I spent an hour trying to get a decent photograph of him. He represented another quest bird; a raspberry-thicket specialist, the Mourning Warbler is most common in scrubby, regenerating clearings left after logging or fire. This big and handsome species is a favorite of birders, though not quite so revered as its lookalike, the Connecticut Warbler. The Mourning is neatly plumaged: dark-gray hood, olive back and wings, and yellow breast and belly. Its species name refers to the blackish bib that hints the bird is dressed for a funeral. A shy thicket-skulker, it is difficult to glimpse, but its strident song gives away its presence: *cheerEE cheerEE cheerEE CHEEReeoh!* This vocalization is evidently a favorite of U.S. TV advertising executives; one can often hear the Mourning's song in make-believe suburban backyards on TV. Few people's backyards actually boast singing Mourning Warblers, but my current side yard in northern Wisconsin did—one of the treats of staying at this little reserve on the Saint Croix.

After dinner, I returned to Crex Meadows to watch the sun set over the marsh. Spring Peepers and Gray Treefrogs chorused. Two American Bitterns called back and forth. At 8:20 p.m., two Sandhill Cranes bugled. Various Sora rails sang out as the sun dropped below the horizon. The light dimmed and the temperature dropped, and yet the male Red-winged Blackbirds continued to vocalize from their exposed call perches, their testosterone pumping. Returning to camp after dark, I was greeted by the rapid and cadenced musical notes of a Whip-poor-will: *wir-wuh-WRILL . . . wir-wuh-WRILL . . . wir-wuh-WRILL.* A bit later, three Barred Owls held a noisy discussion in the trees behind my tent.

On Sunday, I was up well before dawn to search again for that elusive threesome, the Yellow Rail, Le Conte's Sparrow, and Nelson's Sparrow. The cool, misty spring morning was perfect in every way but for the absence of the target birds. Yet other treats—Vesper Sparrow, Red-breasted Nuthatch, Rose-breasted Grosbeak, Sedge Wren—alleviated my disappointment. The Sora rails were singing. A Wilson's Snipe noisily zigzagged past. I saw more large wolf tracks in the sand. A Red Fox crossed the road in front of my car; a flock of Black Terns foraged low over one of the impoundments; and I found a Pine Warbler in a big stand of planted pines. A morning highlight was a singing male Sedge Wren that refused to leave his perch in spite of my close approach; typically, getting a glimpse of this bird verges on the impossible. I spent more than ten minutes admiring the little extrovert. Then a statuesque Sandhill Crane posed for me, allowing me to photograph him from various angles.

I also spotted the Golden-winged Warbler, a quest bird and one of North American birders' rarely seen wood warblers. Once merely uncommon and patchily distributed, its population has declined substantially, and now it is a true rarity. The species breeds only in the mountain uplands of the Appalachians and the North Woods in the northern tier of central states into Manitoba. It favors early successional woodland clearings and bog edges and tends to disappear from places where it has long bred as the vegetation matures. The male is a stunner, with a black throat, black mask, golden cap and wing-bars, and natty gray-and-white body plumage. Seeing this bird sing its brief buzzy song from a small sapling in the sun is one of warbler-watching's high points.

In the afternoon, back in the state park, I biked the Wood River Trail down to Raspberry Landing, on the Saint Croix River. Young secondary hardwood forest, noisy with Veeries and Ovenbirds, dominated the whole area. The White Trillium was flowering in profusion in the forested bottoms, but here in the northlands it was still on the early side of spring. The Black Cherries were just now flowering. The only butterfly I encountered was a Pink-edged Sulphur.

I returned to the Crex Meadows visitor center to get a map, and during our conversation the ranger told me that the town of Grantsburg had trapped and removed twelve nuisance bears the preceding summer. They were translocated to Chippewa National Forest, in northern Minnesota. The presence of bears and wolves seems to be taken in stride in northwestern Wisconsin, and people appreciate them as a part of the local environment. The nuisance animals are managed nonlethally, with a minimum of fuss. In the interior West, where cattle and sheep are grazed on open range, these predators are treated with less forbearance.

In the evening, at 7:42 p.m., a Barred Owl hooted down in the hollow. A gorgeous sunset flamed behind the trees across the Saint Croix River. At 8:20 p.m., a Veery, Ovenbird, and Chestnut-sided Warbler were all still singing. As we head toward the summer solstice, the song of birds extends later and later into the evening hours—a welcome feature of late spring for birders.

A review of the breeding ranges of the boreal wood warblers in the Sibley field guide shows a common pattern of geographic distribution. The breeding ranges of twenty-two species of wood warblers extend from the Canadian Maritimes, New England, and the Great Lakes region northwestward through central Canada to Alaska, with the largest part of the ranges located in eastern, central, and western Canada. This shared breeding range is typified by the ranges of the Tennessee, Cape May, Bay-breasted, and Mourning Warblers. These ranges all have the same environmental theme: they center on the Great North Woods, the spruce-fir forests of the taiga zone. At Crex Meadows, I was at the southern verge of this great swath of boreal vegetation, so extensive in Canada. Many of these warbler distributions also have the same geographic theme: they center on the heartland of northern Ontario.

This is why I was headed to northern Ontario: to visit ground zero for breeding wood warblers. It was a place I had never been, and it is a wild region that has been little surveyed for birds. I did not know what I would find, but I had hopes that I'd see *lots and lots* of wood warblers of many species singing on their breeding territories. But, before my visit to Ontario, I had three stops to make in Minnesota.

REVERSE MIGRATION

The Tuesday after Memorial Day breaks rain free but misty. A stuttering Hairy Woodpecker and a singing Scarlet Tanager send me on my way to Duluth, Minnesota. I take the long way, visiting the Barnes and Mokwah pine barrens in search of patches of young Jack Pine and the rare Kirtland's Warbler. I find plenty of pines, some decent Kirtland's habitat manufactured by the state, but no warblers. I also see along the way that northern Wisconsin, cut over repeatedly, remains a land of timber extraction and forest management focused on the timber and pulp industry—and not much on wilderness. Virtually all the roadside habitat I see is young second growth.

In the early afternoon, I cross the big bridge from Superior, Wisconsin, to Duluth, Minnesota. On the south side of town I look up a hill to a small local ski area, where a snow patch lingers on a ski run—a remnant of winter.

I'd scheduled a meeting with expert local birders Larry and Jan Kraemer at their home on a hill above Duluth this afternoon. The

Kraemers know the area's birds and wildlife and agreed to advise me on birding in the North Woods. We chatted about birding while we watched the bird feeders on their back porch: Pine Siskins, American Goldfinches, and various other songbirds came in to entertain us. The Kraemers noted that they had recorded 161 species of birds in their yard, including twenty-seven species of warblers, along with Black Bear, Coyote, Bobcat, both species of foxes, Pine Marten, and Fisher—and all this in suburban Duluth.

The Kraemers also related the story of a major *reverse* songbird migration, which took place on May 19, 2013, on the southwest shore of Lake Superior at Duluth. During inclement weather accompanied by strong northeasterly winds, tens of thousands of songbirds that had already passed Duluth on their way north reversed course and returned *southward* in search of better weather and foraging opportunities. It was estimated that more than 80 percent of the southward-moving songbirds seen were not identified because of the terrific winds, which made it nearly impossible to focus binoculars on the flying birds. It was the biggest local fall-out of thrushes, warblers, and other passerines in recent memory.

What they reported, however, was but a tip of an avian iceberg. None of the larger trees and shrubs on the shore of Lake Superior were leafed out due to winterlike weather that had extended well into May. Duluth had had its snowiest April ever, with more than fifty inches of snow and persistent cold. Despite the strong offshore winds, hundreds (perhaps thousands) of warblers were desperately trying to find food and shelter among the grasses and small shrubs along the lakeside dunes. Hummingbirds tried to find sustenance from willow catkins and buds on fruiting trees, without much success. Warblers congregated to forage along the southern shore of Lake Superior, Northern Waterthrush, American Redstart, and Magnolia and Yellow Warblers common among them. Orange-crowned, Tennessee, and Cape May Warblers were among the most common species foraging at or below eye level in willow, Red Osier Dogwood, and other small shrubs. American Redstarts and Cape May, Magnolia, and Chestnut-sided Warblers foraged on the ground and chased aerial insects from low perches. Warblers searched for invertebrate prey and any other available sustenance in the detritus washed up on the beach. Hundreds of Palm Warblers foraged along the wrack line. Most surprising to the observers were the Blackburnian and Blackpoll Warblers (species normally seen foraging in the high canopy) and the Mourning and Canada Warblers (species usually seen skulking in heavy undergrowth) that were out in the open, picking at debris down on the beach. Though adding a splash of color to the shore on a dreary and overcast day, the observers realized that these birds were suffering an existential threat from the cold, wind, and lack of food.

During the fall-out, the Kraemers, along with local birding colleagues Mike Hendrickson and Peder Svingen, made observations at several sites around Duluth and the lakefront, and Svingen compiled a full report (which I relied on to retell the story here). The birders also noted an additional *four thousand* warblers that they were unable to identify to species. This incident of reverse migration in spring was new to me. Given the meteorological conditions, I hope I never experience one, for the sake of the birdlife. Still, it is a remarkable phenomenon that once again reveals the courage and tenacity of songbird migrants even under perilous climatological conditions. Presumably most of these birds survived the terrible storm (the observers found few dead birds in its aftermath), but certainly there must have been some mortality.

HEADWATERS

After a night's stay in Duluth, I head westward to Park Rapids, in central Minnesota. Whereas Duluth, in eastern Minnesota, is deep in the forest zone, the western half of Minnesota is prairie country. The town of Park Rapids is right on the boundary between forest and prairie, where farmers grow corn, soybeans, and potatoes. It lies just south of Itasca State Park, which hosts the headwaters of the Mississippi River—one focus of my northward journey's route for the past two months. I have been following the river's course for weeks, and now I am almost at its source. Ornithologist Marshall Howe and his wife, Janet McMillen, live in Park Rapids, and Howe has offered to take me birding at the headwaters.

The three-hour drive from Duluth was punctuated by a close encounter with a Sandhill Crane family foraging at the very edge of the highway. In front of me, a large truck roared by the group, and its slipstream toppled the two rusty-colored and fuzzy young into a grassy ditch. The parent cranes appeared nonchalant about this. When I stopped to look at the cranes, I saw Bobolinks in the field beside the road. The male Bobolink is mainly black, with bold white patches on the wings and a pale yellow nape, whereas the female has the look of a large, plain sparrow. Male Bobolinks displaying over an old field is an iconic boreal spring sight, made richer by the exuberant bubbling song of the displaying birds as they hover over the tall grass.

In midafternoon I booked into Breezy Pines Campground, a few miles from Howe and McMillen's home and hidden in the woods at the edge of small Crooked Lake. My campsite was idyllic—isolated from the rest of the main campground. From the picnic table I had a lovely lakefront vista west to north-country sunsets. And the birdlife was great, too: I hadn't expected to see so many southern species so far north, but here in north-central Minnesota I would find Warbling Vireo, Baltimore Oriole, and Great Crested Flycatcher, among others. Still, I *was* up north, as the Common Loons' song each night on the lake would remind me. A regal-looking adult Bald Eagle liked to perch atop a tall White Pine across the lake from the campsite, and many other species were in song—Red-winged Blackbirds, Common Yellowthroats, Yellow Warblers, Veeries, and American Robins.

On the first day of June, Howe took me birding. First we visited Pine Lake County Forest, where we found a breeding male Golden-winged Warbler in song. I had seen the species several times at Crex Meadows, but one never tires of spending time with this rare bird. Then we moved on to Lake Alice Bog for boreal specialties, finding

Yellow-bellied Flycatcher, Northern Waterthrush, Black-throated Green Warbler, and a singing Lincoln's Sparrow.

But the day's main focus was Itasca State Park. We walked the spruce-forested Beaver Trail on the back side of the park, where we came upon a dried carcass of a long-dead Timber Wolf, completely intact and with a horrible toothy grimace. I couldn't take my eyes off the grisly sight. We wondered what happened to the poor beast: perhaps it had died at the height of winter and been naturally freeze-dried. Not much farther down the trail lay the severed leg of a Snow-shoe Hare. The name of the trail, I thought, perhaps should be "Red in Tooth and Claw."

Then Howe pulled one of his hearing aids from his ear and told me to pop it into my own. Suddenly a new soundscape appeared. I could hear several high-pitched warbler songs: there was a Blackbur-nian! Aha, a Black-and-white! A Canada Warbler in that dark forest thicket! Over there—a Golden-crowned Kinglet! It was a revelation: here was a device that could bring back much of the hearing of my youth. I high-fived Howe and, of course, refused to return the hearing aid until the end of the day. Now I was angry at myself for not pur-chasing a pair of hearing aids prior to the start of this journey; I hadn't known what I was missing.

Our morning in and around Itasca produced several new quest warblers. The Northern Waterthrush lies at the low end of the beauty spectrum but still was new to my quest. Another of the sparrowlike warblers, this species is dark olive-brown above, dull whitish-buff below, with dark breast striping and a pale eyebrow. It forages on the ground in bogs and other northern wetlands, breeds from New England to Alaska, and winters south to Ecuador. Its crowning glory is its super-loud chattering song, which carries far across the boreal landscape—the most explosive song of any warbler.

The Blackburnian Warbler, a common conifer-loving species and another quest bird, is one of the most admired wood warblers because of the male's colors and patterning: glowing-orange throat and cheek and black-patterned head, back, and flanks, with streaks and splotches of bright white in all the right places. This gorgeous canopy dweller, which winters from northern South America into the Andes, has a very high-pitched, lisping song and is easily missed by novice birders standing far below if they don't crane their necks to see into the tops of hemlocks and spruces. It had been a long time since I had been able to hear this bird sing, but today, with the loan of Howe's hearing aid, I could.

A third quest bird was the Black-throated Green Warbler. One of most familiar boreal warblers of the East, it ranges along the Appalachians south to Georgia and is most commonplace in the mixed conifer forests of New England and Canada. Wintering south to Colombia and Venezuela, it tends to be one of the earliest warbler migrants to return northward, heralding the arrival of the warbler waves. The male sports a black bib, yellow cheek, and green crown, back, and rump, and the plumage is further enlivened by black-and-white flanks and white wing-bars. Its cheery, buzzy, musical song is an announcement of spring migration: *zee zee zee zo zeeet!*

We heard a fourth quest species singing from its territory in a dark glade: the Canada Warbler, another of the boreal breeders, with a nesting range in the Appalachians, New England, and Canada. Wintering in the Andes and northern South America, it is an uncommon migrant and tends to lurk in low tangles in mixed northern forest during the nesting season. This is one of the more handsome species, with delicate patterning: gray above, yellow below, with a black necklace and a black face with yellow "spectacles." Its song—high, rapid, and chattery—is a distinctive voice of the dark forest interior.

The morning, with four new warblers, was a substantial step forward for my breeding warbler list, bringing me to twenty-four species. After a rewarding morning of warbler watching, we lunched in Itasca Park at Douglas Lodge, a handsome 1905 log structure with a huge stone chimney and fireplace, a perfect place to relax after a stiff session of wilderness birding. We relaxed, gazed out the window at the greenery, and dined slowly, savoring our meal and recalling the beautiful birds we had encountered.

Blackburnian Warbler

After we got our strength back, we drove a short distance to the Mississippi Headwaters Center and located the trailhead to the headwaters track for our pilgrimage to the source of the Mississippi. There is a lot of hype surrounding the source of "the Father of Waters," but probably, for most pilgrims, it ends in disappointment; the flat path to the headwaters, no more than a couple of hundred yards from the crowded parking lot, ends at an underwhelming small, shallow stream flowing out of Lake Itasca. We learned that the Mississippi starts as this inconsequential feature on the landscape: there's no waterfall here, no mountain tarn, just the small outlet stream. Howe and I were among a crowd of tourists shuffling down the short headwaters trail; when they arrived at the signboard identifying the source of the Mississippi, most snapped a quick picture and then turned to head back to the parking lot. Not all headwaters are so disappointing, of course—the headwaters of the Connecticut and Hudson rivers, for example, start in beautiful and isolated wilderness spots high in the mountains.

The highlight of this iconic geographic locus was in fact a bird—a glossy adult male Black-backed Woodpecker, a rare boreal forest species, which flew into a big White Spruce next to a gravel path near the visitor center and offered us stupendously close looks before disappearing into the woods. It was our seventh species of woodpecker that day. Families and groups of tourists passed us by, oblivious to the elusive and handsome woodpecker. We also found a colony of a beautiful wildflower, the Large-flowered Bellwort, whose leaves look a bit like those of Solomon's Seal, and whose bell-shaped yellow flowers hang off their stems. Returning to Howe's home just outside Park Rapids, we found a wooded section of back road that swarmed with Band-winged Meadowhawks, which are large dragonflies with a bronze abdomen. Hundreds of the large odonates zipped back and forth over the road in an explosive emergence. They, too, were among the day's best sightings, and more of spring's riches.

WARBLER SONG AND TERRITORIALITY

Aside from their remarkable migratory story, warblers attract attention because of the males' brightly patterned plumages and their diverse and often complex songs. The vivid plumage and song relate to the males' territorial behavior: a male warbler arrives on his territory in late spring and spends much of each day for several weeks advertising and defending it. He sings thousands of times each day on territory to announce his presence to competitors, and his declaration of territorial ownership is a way that he defends his patch from other males. In addition, he employs the song to attract a mate to his territory. While males pursue this spring period of territorial establishment, they vocalize and perch in prominent places and make themselves accessible to birdwatchers. It is a behavior of such vitality and vigor that it brings joy to the heart of any birder experiencing it. When territorial song is being deployed by literally hundreds of birds scattered through the woods, it is one of nature's most compelling seasonal events, and it is one of the great attractions of the boreal North Woods to naturalists and birders. Of course, we also love wood warblers because of their favored habitat: mature forest that in spring bristles with life of all kinds.

MANUFACTURING HABITAT FOR THE GOLDENWING

The morning after my insider's tour of the Mississippi headwaters environs, I say my farewells to Howe and McMillen, break camp early, and depart toward forty-three-thousand-acre Tamarac National Wildlife Refuge, an hour west on back roads, where I'll visit an American Bird Conservancy field project. Situated where the tallgrass prairie meets boreal and northern hardwood forests, Tamarac's great mix of lakes, bogs, wetlands, streams, and forests is home to healthy populations of Timber Wolves, Trumpeter Swans, and Bald Eagles, as well as scores of breeding migratory songbirds. On the drive, I pass a pure stand of oaks that still has tiny, very pale young leaves. Spring is slow getting to some of these places.

Meeting Peter Dieser and Aditi Desai of the American Bird Conservancy at the refuge headquarters, I headed out into the field with them to see what the conservancy is doing to increase Tamarac's breeding habitat for Golden-winged Warbler and American Woodcock, both of which have been in widespread decline and are on the Watch List of the North American Bird Conservation Initiative (the NABCI, as readers will recall, hosts the Joint Ventures work that I visited in Missouri). The woodcock is a chubby, short-legged, long-billed, bug-eyed sandpiper relative that has evolved to inhabit wet deciduous woods rather than the seashore. It breeds in the eastern United States and winters in the Southeast. Birders head out to boggy openings in rural areas in early spring to witness the woodcock display flight at dusk, and hunters and their dogs pursue woodcock as a game bird in the autumn.

Dieser led the way to three sites where the refuge, working with ABC, had conducted habitat management plans to open up patches of closed deciduous forest to create early successional glades, prime habitat for Golden-winged Warbler and American Woodcock. This kind of management, as I had seen in Illinois and Missouri, is not for the faint of heart. Heavy equipment is brought in to thin closed forest and enable patchy regrowth of sapling-stage woody vegetation—ideal breeding habitat for the two target species. Although mechanical opening of the forest looks brutal, within a couple of years these areas green up nicely and start to attract the target bird species. All three managed sites at Tamarac were already supporting populations of the two species.

Over the past seven years, ABC, partnering with various private, state, and federal landowners, has created more than twenty thousand acres of new early successional habitat for breeding Golden-winged Warblers and American Woodcocks in New Jersey, Pennsylvania, Maryland, and Wisconsin as well as Minnesota. This is a working experiment, and associated censuses monitor impacts on the abundance of singing males onsite. ABC has found that the target species are already on territory and breeding in a number of the newly managed sites. In addition, recent field studies have shown that early successional habitat is an important foraging resource for young songbirds of the year in the late summer, prior to fall migration. Thus this extensive field experiment may produce additional benefits for migratory songbirds.

At the third site, we met Earl Johnson, a forester and woodcock expert, who was using hunting dogs to census woodcock and locate woodcock nests in an attempt to document the effect of the ongoing habitat management. He showed us a nest, set inconspicuously on the ground in thick grass, with eggs upon which an adult American Woodcock sat. The dorsal plumage of this marvelously camouflaged species is patterned like dead leaves on the forest floor, and even from a few feet away, I'd been unable to detect the bird on its nest.

Weather at Tamarac was gloomy, with on-and-off light rain and mist, but we were pleased that we were still able to carry out the field visit. Aside from the woodcock and Goldenwings, we observed Ring-necked Duck, Black-billed Cuckoo, Veery, Black-and-white Warbler, Yellow-throated Vireo, Scarlet Tanager, and Rose-breasted Grosbeak. I spotted two annoying species as well: Poison Ivy, surprising this far north, and Wood Tick, dozens of which I picked off my clothing at the end of the field visit. Because of warmer winters in recent decades, ticks of several species now abound in north-central Minnesota, as do the several diseases typically associated with tick bites.

Once I dispensed with the ticks, I departed Tamarac and headed northeastward toward International Falls and the border with Canada. To date, I'd tallied twenty-four wood warbler species on their breeding grounds. Thirteen more to go. I was now in a northern zone where singing breeders outnumbered passage migrants, and I had high hopes for northern Ontario.

SEVEN

The Mysterious Northlands
Early to Mid-June 2015

June comes with its own tranquility, predictable as sunrise, reassuring as the coolness of dusk. . . . There is a certainty, an undiminished truth, in sunlight and rain and the fertility of the seed. The fundamentals persist.

—HAL BORLAND, *Sundial of the Seasons*

I break camp in Bemidji, Minnesota, and drive northeastward to the Cana-
dian border. The narrow, empty road is flat and straight and cuts through
clumps of aspens, stands of Balsam Fir and White Spruce, and Tamarack
bogs. Some aspens exhibit young pastel-yellow leaves that seem to shout,
"Early spring!" As the earth continues to tilt slowly southward toward the
sun, I drive north, in retreat from approaching summer.

At the International Falls border crossing I am nervous and jumpy,
but the Canadian guard offers a warm welcome. Crossing the small bridge
over Rainy River to Fort Frances, I am now safely into Canada. I have
five hundred miles of back roads before me to reach the wilds of interior
Ontario—the land of Woodland Caribou, Spruce Grouse, and Wolverine,
and the breeding heartland of the boreal wood warblers. I cruise northward
across Canadian shield rock cloaked with conifer and mixed forest. A large
American Black Bear lumbers across the road in front of my car. Road signs
warn of Moose wherever I cross a wetlands.

oaded down with several weeks' worth of provisions I'd bought in
Dryden, Ontario, I headed northeast to Sioux Lookout and Savant
Lake, into the Kenora District of Ontario. As I drove the lonely
roads, I received only a single signal from my radio: Canadian Broad-
cast Corporation radio (analogous to National Public Radio). I listened
to a discussion regarding a recently released report by the Truth and
Reconciliation Commission, which had issued a formal apology for
the Canadian government's practice of separating First Nations (indigenous) children from their families and placing them in distant "residential schools," cruel treatment experienced by seven generations of
First Nations children. The stated purpose was to train First Nations
children to become "good Canadian citizens." Similar stories abound
about Indigenous Australian children and Native American children
in the United States. As I headed deep into First Nations territory, the
report introduced me to the types of social issues confronting minority
aboriginal people living in white Canada.

OPPOSITE: Magnolia Warbler

My destination today was Pickle Lake, the northernmost community accessible by paved all-weather road in Ontario. As I drove northward, the land became hilly, interspersed with lakes aligned in a northeast-to-southwest orientation. Not so long ago, in geological terms, a great block of ice had trended southwestward over this mass of shield rock, and as it retreated, it left lakes, outwash sand barrens of glacial till, scoured and scored bedrock, and moraines of gravel in its wake. I was now in the middle of a vast territory with few people. Ontario is half again larger than Texas, but most of the province is roadless, and industrial timber extraction has not yet made its way this far north. The province has a quarter-million lakes—*twenty-five* times the number in Minnesota (the land o' lakes).

WALLEYE NATION

Moving northward, I passed the occasional sign pointing to road access to a resort fishing camp. It turns out that this is major recreational fishing country: the hundreds of lakes scattered east and west of Route 599 and its extension, the gravel-only Nord Road, support hungry populations of two important northern game fish, Northern Pike and Walleye. The elongated, sharp-toothed pike, a large predaceous fish with a Northern Hemisphere distribution, typically maxes out at about three feet long and twenty pounds. It is a popular game fish but is not prized for eating in North America because it is very bony. The smaller Walleye is a North American perch, growing to eighteen inches and weighing a couple of pounds. It is a beloved game and eating fish for Canadians as well as Americans living in the northern Midwest; fishing for it is a big deal from northern Ohio to Minnesota. In northern Ontario, Walleye (called Pickerel by Canadians) is the main sport-fishing target.

The road north was empty of traffic, and I saw no one out and about in the several newly constructed First Nations settlements along the roadside. In fact, most First Nations settlements here lie off the highway, deep in the woods. Because of the boggy landscape, many of them are accessible by SUV or truck only in winter, on the "ice

North on the Wing

roads" made famous by the American cable TV show named for them. At about 7:30 p.m., I passed the community of New Osnaburgh, the Anglo name for the tribal seat of the Mishkeegogamang Band of the Ojibway Nation, and at last, after thirteen hours of driving, I arrived at Pickle Lake. This is where the paved road ends. In earlier decades, the town was bustling, enriched by local gold mining. Now it is primarily the site of local government and health facilities serving the vast First Nations territory that stretches to the horizon in all directions.

LAND OF THE MISHKEEGOGAMANG

In Pickle Lake, First Nations people stand on the street corners or gather in small groups. Few cars pass; most people walk to get where they need to go. Something seems lacking in the town. It has a mournful feeling, even though the sky is blue and the sun is shining. This was once an Anglo boom town, but those days are gone, and now the indigenous community seems weighed down by existential angst. I rest for the night in a guest house's backyard in town, and in the morning break down my tent in the frosty morning air, pick up a camping permit from the Government Office, and buy gas from a First Nations convenience store. I head up the Nord Road to camp for a couple of weeks at three North Woods sites: Menako Lakes, Pipestone, and Badesdawa Lake, all in the Kenora District and all offering productive Walleye fishing. However, few birdwatchers like me venture this far north.

The seventy-mile drive north on the wide gravel road wound around bogs, lakes, ridges, and rivers. Small First Nations camps sat just off the roadside: the Mishkeegogamang, enthusiastic Moose hunters, use the camps for autumn and winter hunting, trapping, fishing, and gathering. Even though most of this area is Crown Land, First Nations people, based out of two small reservations, claim the land as traditional territory and are permitted to hunt and trap on it.

The lake and river waters were cold and dark with tannins. The district campgrounds were spartan but nicely situated on the water,

American Three-toed Woodpecker

with places to put in a kayak or canoe. I set up my tent at Menako Lakes campsite, stretching a big nylon tarpaulin over my tent and the adjacent picnic table that would serve as my kitchen and dining room. Black flies and moose flies were bad during the day here, and mosquitoes swarmed in the evenings. To counteract the biting insects, I broke out a novel insect repellent device called Thermacell, a plastic unit about the size of a cordless phone handset. It uses a small butane cartridge to heat a small square fiber pad impregnated with synthetic pyrethrum. The heat vaporizes the repellent and creates an invisible protective cloud. By running the device on my picnic table, I could create a bug-free zone under the canopy of the rain tarp, which made cooking and dining up here bearable. Without it, I would have had to dine with a head net and gloves.

Now I needed to get to know my new environment. I went out on bike trips mornings and afternoons, covering a lot of territory. I used the car for more distant travel in search of big game or unusual habitats, and drove to Musselwhite Gold Mine, about forty miles distant, to purchase gasoline and to get ice and supplies for the cooler. The mine, an isolated industrial operation, sat at the end of a long private access road northeast of Pipestone campground. I spent evenings kayaking to watch wildlife and to escape the mosquitoes, which followed my kayak out quite a distance, giving up only after I'd traveled a couple hundred yards from shore. Out on the water, I drifted and watched the sun setting behind the low hills, satisfied to be in wild nature. Spring Peepers chorused loudly and a Canadian Toad trilled. This was a superior way to end the day. The sun hit the horizon at a bit before ten, so I was in my tent before dark on most nights. My sixteen days up here in the far north followed this general routine: several sallies out on foot, by bike, or in the car to hunt for birds and other wildlife, interspersed by simple meals, the occasional afternoon siesta, and a quiet and very peaceful night of sleep in the tent, only to be woken early by birdsong.

BIRDING THE JACK PINE BARRENS

Much of the upland territory where I camped was forested in a monoculture of Jack Pine. A large stretch of northern Ontario is considered a "barren," underlain by nutrient-poor shield rock. The only soil available since the glaciers receded eleven thousand years ago is mostly sand. Low spots are boggy and wet, but the dominant uplands are dry and sparsely vegetated with the fire-prone Jack Pine. The summer brings drought conditions, when lightning strikes produce fires that, under these desiccated conditions, can rage for days. Jack Pine, with its sappy resins, burns like crazy, but this species happens to be a fire specialist, its seeds released from its cones by a fire event. Once I walked through the edge of a burned-over section of Jack Pine forest and saw, among the blackened and charred trunks, thousands of tiny green seedlings—Jack Pine begetting Jack Pine.

The dominant dry upland forest here has an overstory of mature Jack Pine, with a scattering of fir and spruce saplings in the understory. Here a mature Jack Pine might stand thirty-five feet tall, with a basal diameter of no more than four inches. This, then, is "old-growth" Jack Pine, but I wondered how old it actually is—fifty years? The Jack Pine stands are quite open, and the ground is spongy, with a thick carpet of *Sphagnum* moss and Reindeer Lichen. In certain spots, spruce, fir, aspens, and poplars prosper, but Jack Pine is the dominant tree in this part of the North Woods.

So, the first jolting discovery of my North Woods sojourn was that my patch was not what I thought it would be. I'd assumed I would be in the heart of a mature spruce-fir forest, but instead I was in a monoculture of scrubby and fire-prone Jack Pine. Based on research I had done at the National Fish and Wildlife Foundation, I knew Jack Pine was a strange and stunted little conifer required as breeding habitat for the endangered Kirtland's Warbler. I had driven earlier through Jack Pine barrens in northern Wisconsin, and had understood why the pine gets no respect: it is short and often irregular in shape. It is a poor timber tree. And it tends to burn up in forest fires before growing very large.

This barren landscape was now my temporary home, and I ventured out farther from my campsite as the days passed and the summer solstice drew near, giving me more than sixteen hours a day of daylight: a lot of time for naturizing. And I needed it. The Jack Pines' dominance meant that I had to search for each new animal I saw: few birds sang in this great expanse of piney scrub. The birds that summer here occupy the forest in vanishingly low numbers.

The Ruby-crowned Kinglet was the most common long-distance migrant inhabiting the monoculture: wherever I went, I heard its high, bubbling crazy quilt of a song, one of the few I recorded every day in the North Woods. The energetic little songbird nervously foraged in the needles of the Jack Pines, usually on upper branches, with flicking wings and herky-jerky movements.

Second only to the kinglet in the pines was the Gray Jay, an iconic bird of the North Woods, which I also recorded each day. This nonmigratory, year-round resident is very different from the more familiar Blue Jay, which does not range this far north. The Blue is migratory, quite vocal, and wary, whereas the Gray is nonmigratory, quiet, and tame. The very plain Gray Jay somehow manages to survive the brutal winters and nest productively in the North Woods. Adults form pairs that stay together year-round. Nesting in early spring, they produce two or three blackish young that hang out with their parents for much of the late spring and summer. Family groups of Grays emerged from the forest to investigate me as I passed by on my bicycle, showing no fear and allowing a close approach. It is hard not to like these whimsical and confiding birds in this lonely land.

Three other birds do fairly well in this Jack Pine landscape: the Common Raven, American Robin, and Northern Flicker. I observed the three most days when I cruised the Nord Road through the vast stands of Jack Pine. The raven is our largest songbird, a year-round resident of the North Woods, and it is one of those birds that moves far and wide rather than being a habitat specialist; it tolerates the Jack Pine landscape but can use many other habitats to survive. Here the ravens moved up and down the Nord Road in search of roadkill and other carrion, and the opportunists were hyperabundant at the Musselwhite Mine, where flocks of the big black birds hunkered on every available perch, reminding me of ominous scenes in Hitchcock's *The Birds*.

The robin, that familiar backyard songbird, loves wilderness as much as suburban neighborhoods, and I found it in small but decent numbers in the Jack Pines. Here in the North Woods, it is a summer visitor; these northern populations winter south into the United States. The bird has very broad habitat tolerances and, like the raven, is able to make a living in just about any wooded environment, which makes it one of North America's most successful songbirds.

I spotted the Northern Flicker in small numbers in the Jack Pines; this woodpecker is another commonplace North American species that breeds from Alaska to south Florida. As with the robin, the flicker is a summer breeding visitor to the North Woods, here foraging in the clearings created by the Nord Road. It winters south into the Lower Forty-eight.

During my stay in the North Woods, I occasionally encountered other birds in the Jack Pine monoculture—Myrtle Warbler, Tennessee

Warbler, Golden-crowned Kinglet, Red-breasted Nuthatch, Ovenbird, Hermit Thrush, and Dark-eyed Junco—but most of them were more common in other wooded habitats. The Myrtle Warbler, alternatively known as the Yellow-rumped Warbler, lives in the northern spruce bogs as well as the pine barrens, and it became the first quest species I saw in Ontario. The species is an annoyingly common passage migrant in spring and fall in the Mid-Atlantic: flocks of this bird can swamp out other warblers during migration. Moreover, many Myrtle Warblers winter along the East Coast in coastal scrub, where they are one of the more common species of birds in the dark of winter. They are one of the few warblers to subsist on fruit in the U.S. winter, eating the fruit of Bayberries and Wax Myrtle. The male sports a yellow crown-spot, flank patch, and rump, which highlights its black, gray, and white plumage. The species breeds in conifers and sings its feeble musical warble from atop conifer spires.

Yet wherever songbirds are few in a habitat, hearing a Myrtle Warbler in song is reassuring. The Jack Pine forest was an ornithological dry hole, but perhaps I would find my migrant songbirds in the two other habitats: the spruce bogs and the aspen groves.

BIRDING THE SPRUCE BOGS

The patches of forest that featured spruce, fir, and Tamarack proved to be birdier habitat, and I found them mostly in the boreal bogs. I'd known and loved such areas during my years spent in the Adirondacks, where the specialty boreal birds that I knew lived in spruce bogs. It was thus clear that I needed to search out and spend as much time as possible in bogs here in northern Ontario.

A typical spruce bog settles into a depression in the bedrock that traps water. Usually at the bog's center is a pond that is being gradually encircled by a mat of *Sphagnum* moss, which can grow in dark, acidic water. The acidity is created by the chemistry of the parent rock and by the annual deposition of conifer needles. Year after year, the *Sphagnum* adds layers and sinks into the water, forming a flat mat (called peat) that thickens and that creeps ever closer to the center of the pond. Acid-loving shrubs start to colonize the growing mat of *Sphagnum*, and then the bog-loving conifers (Black Spruce, Tamarack) follow, growing on the driest outer perimeter of the *Sphagnum* mat and encircling the pond.

From above, a mature spruce bog reveals itself: the small pond is surrounded by a broad ring of *Sphagnum*, encircled by various bog-loving shrubs (Labrador Tea, Bog Laurel, Sheep Laurel, Dwarf Huckleberry, Leatherleaf), surrounded by rings of ever-larger conifers, the most mature and tallest of them on terra firma, at the outer edge of the depression. Thus a classic spruce bog has an exquisite natural design. Nature then throws in a number of unusual small herbaceous plants—pitcher plants, sundew, and various orchids—just to make things interesting for botanists.

Bogs come in many other shapes, too, not just the typical circular form. They can arise when a receding glacier drops a pile of rock as a moraine that blocks a valley, causing flowing water to fill the depression. In some places, vast planar boglands can form without a central pond. Instead they hold an open *Sphagnum* mat, dotted with tiny spruce and tamarack trees, that stretches across the landscape.

Along the Nord Road, a mix of bogs and boglands is scattered within the Jack Pine landscape. Here the finest boreal conifer forest grows on morainal edges of bogs where the soil drains better and where the trees are protected from the periodic fire that sweeps through the Jack Pines every few decades. The great fires burn patchily, destroying some areas and entirely missing others. Spruce, fir, and Tamarack are not fire-tolerant species, and so these blazes, along with other periodic disturbances and varying groundwater availability, create the patchwork of forest types I saw here. In this area, spruce bogs occupy about 15 percent of the land cover, and thus I had plenty of territory to explore, which I did day after long day.

As noted, the spruce bogs of the North Woods resembled those I knew from the Adirondacks. Because the big ice sheets that trundled over these lands also made their way down to the present site of New York City before retreating, the youthful boreal woodland flora in New

Spruce Grouse

England and northern Ontario are very similar, and both are species-poor. Black Spruce and Tamarack, which populate bogs in the Adirondacks, also populate the bogs in the North Woods.

Here along the Nord Road, Tennessee and Magnolia Warblers sang from tall mature conifers that grew around the bogs' outer verges. Hermit Thrushes and Swainson's Thrushes sang their beautiful songs from the dark recesses of the conifer stands, where Winter Wrens bopped about and launched into their virtuoso song solos from thickets and blowdowns. Golden-crowned Kinglets foraged quietly in the top branches of spruces. White-throated Sparrows hid in shrub thickets at bog edges and sang their sad "Old Sam Peabody" song. The occasional tall, dead spruce snag would host an Olive-sided Flycatcher, singing his *quick-three-beers* song every now and then. It was a wonder to watch this rare flycatcher launch out high into the air in pursuit of some distant flying insect.

Two other, smaller flycatchers frequented the spruce bogs, too. The noisiest was the Alder Flycatcher, which sat atop Speckled Alders in wet spots in the bogs. This drab flycatcher gave its burry *wee-BEE-oh* several times a minute, flicking its wings. The Yellow-bellied Flycatcher, a

North on the Wing

conifer specialist, usually hid in the spruces and firs to give its soft *chee-bunk* or sometimes a liquid *per-wee* call. The bog songbird with the prettiest voice was Lincoln's Sparrow, whose sweet series of various trills carried far and gave away the location of this very shy shrub-dweller, typically seen hiding nervously in the center of some bushes.

The Tennessee Warbler (discussed at length in chapter 1) was one of the quest birds that I found in the spruce bogs. In fact, it was the most common wood warbler here. I saw and heard it daily, and wondered why this species abounded here when other warblers did not. This is a question I never have answered.

Two of the open-bog specialists were Nashville Warbler and Palm Warbler. These handsome yellow-washed wood warblers flitted about in the tops of isolated Tamaracks and spruces scattered across the mats of *Sphagnum*, their loud songs giving them away. The Nashville Warbler is one of the more common and widespread of the boreal wood warblers (we discussed the species back in chapter 2, at the Mad Island banding station). It breeds from eastern Canada and New England west to California and north into Canada. One of the smallest wood warblers, it is a relentless songster, perching atop some conifer in the open and belting out its series of trills and slurs. It prospers in bogs and conifer openings, though it was uncommon here in the far north.

The Palm Warbler, the commonplace vanguard spring migrant along the East Coast, is a true bog specialist—even more so than the Nashville. Eastern populations of the Palm Warbler, which winters from North Carolina to Panama, exhibit a red-brown cap, yellow underparts with red-brown streaks, and an olive back and wings. It exhibits two distinctive habits: tail-wagging and spending a lot of time foraging on the ground in openings in fall and winter. The species rarely makes the top ten most-wanted warblers, mainly because it is a very frequent seasonal migrant throughout most of the United States.

The crown jewel of spruce bog songbirds is the Connecticut Warbler. One morning while biking, I heard the staccato song of a male Connecticut Warbler on territory at the back of a small bog surrounded by tall Tamaracks. I drew him out by playing a recording of the species' song from my iPhone attached to a small speaker, and photographed him while he darted about in the shade. Many birders in the United States say the Connecticut Warbler is one of the most difficult wood warblers to see. It is not terribly rare, but its northern breeding range is out of reach for most, and during migration it is very elusive. I spent years hoping to encounter a Connecticut Warbler in the autumn at Cape May, New Jersey. The closest I'd come was when birding expert Michael O'Brien once called out, "Connecticut Warbler!" while pointing to a tiny bird passing overhead in the wind at Higbee Beach. I could barely even discern that it *was* a warbler at all. But here in Ontario, I had my first good, long look at a singing male Connecticut in full view, unobscured by vegetation.

During the breeding season, the Connecticut Warbler is sparsely distributed in Black Spruce–Tamarack bogs from the boreal zone of northwestern Canada to the Upper Peninsula of Michigan. It winters in northern South America. During spring migration, these furtive birds migrate up through Florida and then west of the Appalachians to the North Woods. In autumn, they move southeast to the New Jersey coast and then out across the Atlantic to northern South America (more on that migratory feat in the last chapter). Most U.S. birders search for this bog specialist in northern portions of Minnesota, Wisconsin, and Michigan, but the hunting is tough: at the southern fringe of its breeding range, this species is few and far between. One aspect of the species life history does give an advantage to the birder: the male's territorial song, a loud and rapid series of *chippy chuppy* notes that carries a considerable distance and is most similar to that of the Northern Waterthrush.

Day after day I hunted the tall, bog-fringing spruces for signs of either Bay-breasted or Cape May Warblers and listened for their high-pitched songs. No luck. Noisy and widespread Tennessee Warblers sang from the tall conifers at the edge of every bog, but many of the expected wood warblers were nowhere to be seen. Though disappointed, I did manage to continue to pick up odds, ends, and surprises. One day a flock of Cedar Waxwings zipped overhead. A few pairs of tannin-stained Sandhill Cranes nested in some of the larger boglands, and their bugling sounded in the distance from time to time. This haunting voice, heard mainly at a far remove, is the song of the wild. In larger bogs, I occasionally flushed out a Wilson's Snipe, a boreal-nesting shorebird that forages in wet grasslands and explodes out of the grass with loudly voiced expletives, causing the flusher to leap backward and gasp. Once on the wing, the bird rapidly zigs and zags across the sky and just as quickly drops back into another patch of boggy grass. One rarely gets a good look at this long-billed, speckled, and striped nonpasserine. And Blue-headed Vireos sometimes appeared in small clearings, foraging atop a conifer.

HOMING RIDDLES FOR THE SONGBIRD MIGRANTS

How did these singing Blue-headed Vireos in the bog edges get to their little patches of Ontario from their winter home in Central America? Just how these songbirds succeed in traveling thousands of miles from their southern haunts to pinpoint their summer nesting site remains mysterious. Since they fly at night and at great altitude, their guidance systems must in some way approximate those of a modern airliner, with the ability to sense various geographic clues to aid their steering. As mentioned earlier in the book, a navigating bird must have analogs to a map, a compass, a calendar, and a clock, as well as a good memory, to make its annual round-trip from the Tropics back up to Canada. Research over the past half-century has enumerated the considerable detection skills that birds use in their navigation to distant fixed destinations. These include the following:

1. the ability to detect compass direction from the position of the sun
2. the ability to detect ultraviolet light
3. the ability to determine the plane of polarized sunlight
4. the ability to detect the earth's magnetic field
5. the ability to hear ultra-low-frequency sound

6. the ability to determine compass direction from the rotation of constellations in the night sky
7. the ability to detect latitudinal position by the seasonal position of particular constellations or through the level of magnetic declination

These, and as-yet undiscovered capacities, enable birds to accomplish their biggest migratory task: locating their nesting sites after eight months away from their northern homes.

Translocation studies show that adult birds of many species, when captured, moved, and then released far from their home territories, are able to make their way back to their nest sites in short order. For instance, an adult Laysan Albatross taken from its nest and released three thousand miles away was able to make its way back to its nest in only ten days—traveling an average of three hundred miles a day. By contrast, young birds are unable to accomplish such a feat. The precise mechanism of target-oriented navigation remains unknown, but certainly it involves the tools listed above and probably additional ones not yet discovered.

How does our Blue-headed Vireo locate the thicket it was born in after a winter in Colombia? Study of different migratory species, thrushes, has shown that once northbound birds reach the latitude of their natal territory, they begin making east-west flights to track down their home. These flights might allow the birds to detect familiar visual landmarks as well as characteristic low-frequency sounds produced by mountains, rivers, and coastlines. The birds also may employ their sense of smell to "sniff" out familiar landscape smells that emanate from the natal territorial (a spruce bog, for example, smells quite different from Jack Pine forest). Using their sharp memory, they might be able to recognize particular landscape features from high in the sky. We know some of the detection tools they deploy, but we do not know which specific tactics the Blue-headed Vireo uses, or how it uses them, to return to its little breeding patch.

BIRDING THE ASPEN-POPLAR GROVES

The third wooded ecosystem here in the northlands is the aspen-poplar grove. This broadleaf habitat prospers in soils of deep gravel on well-drained upland sites. Quaking Aspen and Balsam Poplar—rapid-growing early successional members of the poplar lineage—dominate these groves, which here grow up mainly in the clearings that road crews created when they quarried gravel to build the Nord Road. These patches of mechanical disturbance quickly revegetated in shrubs and the two species of poplars, areas that I spent time in to find songbirds that favor deciduous trees amid the vast conifer barren.

After my success in bringing in the Connecticut Warbler with a playback of its song, I started to use the same technique to search for other elusive songbirds. Since aspen groves are a favored habitat of the rare Philadelphia Vireo, I played its song when I visited the aspens, and I managed to draw in both the Philadelphia *and* the more commonplace Red-eyed Vireo. Earlier in the book I had noted that the songs of the Red-eyed and Philadelphia Vireos are difficult to tell apart, and it seems the birds themselves have this same difficulty. The Red-eyed is the more handsome of the two, with its distinctive dark facial pattern, clear white underparts, and larger bill. The Philadelphia, short-billed and exhibiting a distinctive strong yellow wash on the throat, otherwise looks like an undersized Red-eyed.

Aside from the Philadelphia Vireo, the other true aspen specialist is our familiar Yellow-bellied Sapsucker. The sapsucker, which I first encountered in southern Arkansas, loves to drill its nest hole in a dead aspen stub, and it will drill sap wells in living aspens and poplars. Without aspen groves, the cadenced drumming of this stolid but handsome woodpecker would not be heard in the North Woods.

Given that the aspen groves are early successional habitat that arises shortly after disturbance, these areas include young trees, shrubby thickets, and clearings of bare gravel. Other songbirds like this habitat, and here I added two additional quest species: Orange-crowned and Wilson's Warbler. Both tend to skulk in shrubbery at the edges of clearings, and both signal their presence by their songs, which sound similar. The Wilson's gives a rapid *chi-chi-chi-chu-chu-chu*, whereas the Orange-crowned gives a trilled *ti-ti-ti-ti-ti-tu-tu-tu*. Though distinguishing the songs is a challenge, seeing the birds resolves any uncertainty. The Wilson's is yellow-breasted with a neat black cap. Its breeding range extends from the Eastern Maritimes northwest to Alaska and then south to California and Colorado in the mountains of the West, and it winters from Mexico to Panama. The very plain Orange-crowned is yellowish-olive, with a drab gray head and throat, obscure streaking on the breast and flanks, and a yellowish undertail. The crown-mark for which it is named is never visible to observers in the field. The breeding range of this species generally matches that of the Wilson's Warbler, and its winter range extends from the southern United States to Mexico.

The prize songster of the aspen groves is the rusty-striped Fox Sparrow, which winters south to the Mid-Atlantic, breeds in brushy clearings in the taiga forests of the North Woods, and ranges in summer from Alaska to Labrador. Northern Ontario is on the southern verge of its eastern breeding habitat. This large species is rather easy to observe during the winter, when individuals forage on the ground under backyard feeders. Here in the North Woods, the Fox Sparrow was perhaps the most reclusive songbird: I glimpsed it only once, and a few times heard it singing loudly from the shelter of thickets. The song is a cheerful and complex series of conversational notes, musical and precise. I'd heard it on a few occasions in the Mid-Atlantic in early spring, and I was pleased to hear it now on the bird's breeding ground.

YEAR-ROUND RESIDENT BIRDS OF THE GREAT NORTH WOODS
Two grouse species make this vast woodland their home year-round: the Ruffed Grouse and the Spruce Grouse. The Ruffed has a huge geographic range, extending from Alaska to Labrador and south along the Appalachians as far as Georgia. It prefers mixed coniferous-deciduous woods, whereas the Spruce Grouse sticks to pure conifer stands. The Ruffed Grouse is special because of its nuptial display, called drumming, and I knew the species must be here in the North Woods

because I had heard males drumming at various times during the day and even late at night.

A male Ruffed Grouse attracts females to mate in spring by a set display combining plumage, posturing, and mechanical sound production via percussion. The male locates a fallen hollow log on the ground in a thicket in the woods and determines that the log will suffice as both a display perch and a resonant sounding board. The male attends his display site every day during the mating season. Periodically, he mounts the log and rapidly and rhythmically beats his wings against his ribcage to produce a low-pitched thrumming sound that carries far through the woods. When an interested female approaches, the male does his display, which I watched at close range one morning. It was the first time I had seen a grouse display, and I was mesmerized. The male raised the black ruff around his head, dropped his wings, and lifted and spread his tail feathers in a broad, erect circle. His sound, wing motion, shiny black ruff, and fanned-out tail were a winning combination. He seemed captivated by his own display activity, and I approached quite closely without disturbing the performance.

The little-known Spruce Grouse is a denizen of its namesake forests in the Great North Woods. One of a handful of boreal forest specialists that live in these conifer tracts year-round, it is known informally as the "fool hen," as the species can be inordinately unwary. I found this out at the Pipestone campsite. Biking up the entrance road one morning, I glanced down a side track and saw a big bird standing in the middle of the dusty path. I dropped my bicycle, pulled out my camera, and began slowly edging toward it. It was a female Spruce Grouse, deep brown and heavily barred with fine black markings. It eyed me but did not retreat. When I stood within twenty feet of the bird, I started taking photographs. Before long, I was at the minimum focal distance of the lens: just twelve feet. The bird watched me, pretending to forage from time to time. It eventually flew up into a spruce and observed me from there.

The Spruce Grouse, though insanely tame, is more elusive than the Ruffed Grouse because the male's display includes a short flight, but the wing sound produced does not carry far. I managed to observe

Spruce Grouse on just six occasions during my stay up north, generally detecting its presence only by the distinctive piles of orange extruded tubes of digested vegetable material it excretes, which look a bit like Fiber One cereal. I found these piles in the middle of many sprucey paths. They are a by-product of the species' unusual diet: the needles of spruce and fir. By contrast, the Ruffed Grouse subsists on the buds of ferns and other broad-leaved plants, and I never saw its poop piles on trails or roads.

The Boreal Chickadee and Black-capped Chickadee constitute another pair of permanent residents. These two chickadees are the smallest birds to brave the long, dark winter up north. I encountered the Boreal on four days, the Black-capped on five days—neither species is particularly common here. Black-cappeds range from Jack Pine to spruce bog, but Boreals stay hidden in the spruce thickets. The Boreal differs from the Black-capped in that it is spruce-dependent and has a dull brown cap and reddish-brown flanks. Both are adorable and approachable, but the Black-capped has more personality and is more confiding: it can be fed by hand in winter.

Both species of three-toed woodpeckers are iconic year-round residents of the North Woods. The Black-backed Woodpecker is slightly larger and longer-billed than the less handsome but rarer American Three-toed Woodpecker. Both birds spend all year here, working over dead conifers in search of bark-dwelling and wood-boring insects. One pair of Black-backeds put on a show for me: the birds drummed, and then the male came in excitedly with mewing vocalizations, holding up its wings in a prominent V, perhaps as a sort of territorial declaration. Both species prosper after forest fires because of the standing dead conifers that become infested by a bonanza of wood-boring beetles. The American Three-toed is strictly found in spruce bogs; I saw it just once, while I spotted the Black-backed four times. The Downy, Hairy, and Pileated Woodpeckers—three of North America's more widespread and common woodpeckers—were absent here.

Another North Woods permanent resident is a grail bird for North American birders: the Great Gray Owl. I found a single individual in a large, open bogland area not far from Badesdawa Lake. I played the call of the species after dark, eliciting a very low-pitched series of *hoo* notes from the grand night watcher. Soon I saw the giant owl soar in on set wings, like a gliding moth dark against the pale night sky. He came across a broad expanse of the open bog, almost as if in slow motion, and perched atop a tall conifer, swiveling his great round head. Behind him Venus was low to the horizon, Jupiter higher in the sky. Transfixed by the owl and the shining planets, I barely noticed the penumbra of mosquitoes around my head.

The morning after encountering the Great Gray, I returned to the frost-bedecked bog and tried my luck again. At 5 a.m., I used my iPhone to replay periodically the voice of the big bird. A muffled response came from a thick clump of spruces on the far side of the wetland. I marked the spot and started wading toward it. Twenty minutes later, after clambering over and around numerous blown-down conifers, I was in the thick of the spruces. I scanned for several minutes and spotted the owl, perched about twenty feet up on a large spruce bough in the deep shadows. I was no more than seventy-five feet from him. I crept closer and closer until he filled the viewfinder of my camera. Not at all frightened by my presence, he looked right down at me. His body plumage matched the mottled dark gray of the surrounding spruce trunks, and his big gray eye-disks made his yellow-eyed stare all the more intense.

BIRDS OF THE LAKES AND RIVERS

Quite a few species of birds are closely associated with the abundant lakes and rivers of the North Woods. The most familiar waterbird is the Common Loon, and every lake up here had its pair. They wailed their unearthly banshee yodels on and off through the nights and on foggy mornings. These master fish-hunters like the solitude of unde-veloped lakes and generally don't take kindly to people. On several evenings, I kayaked close to a single adult bird to look at its patterned back and its sleek, flawless neck plumage checkered in black and white. The black bill was sharply pointed and glossy, the eye a deep red. The bird had the look of a powerful submarine. Of course, these birds spend much of their day under water chasing small fish.

One day, out on Middle Menako Lake, I came upon six White Pelicans in their breeding finery who permitted me to kayak right up to them. Then they slowly rose off the water and circled in stately formation, showing me their long black-and-white wings, white bodies, and orange feet and bills. This is the largest bird inhabiting the North Woods, weighing fifteen pounds and boasting a nine-foot wingspan. Their grace in flight belied their ponderous mass.

Many evenings I encountered pairs of Bonaparte's Gulls foraging over the water, hawking for insects much as I had seen Ring-billed Gulls do along the Potomac in Virginia. The petite Bonaparte's is a fairly common winter visitor to the East Coast, but in the nonbreeding season the species' plumage is drab. On its breeding ground, this bird is among the prettiest of gulls. I photographed a pair on a small beach on the Pipestone River to capture the perfect patterning that nature has created: bright-red legs and feet, jet-black bill matching the all-black head, and a white crescent surrounding the dark iris. Their bodies are patterned in snow white, dove gray, and black. In flight, their wings show large white patches bordered by black.

Of course, the North Woods has its breeding waterfowl as well. On evening kayak trips, I encountered pairs of Green-winged Teal, Red-breasted Merganser, Bufflehead, Ring-necked Duck, and Common Goldeneye. This lake-splattered landscape is a magnificent breeding ground for ducks, offering them both ample water and abundant privacy.

NORTH WOODS WILDLIFE AND PLANT LIFE

I was here in search of birds, but I took note of everything else. Butterflies moved through openings in small numbers, and near Menako Lakes I found the Canadian Tiger Swallowtail, the Silvery Blue, a Frigga Fritillary, and a Red-disked Alpine. In other places I found scores of the now-familiar Band-winged Meadowhawk dragonflies active in clearings; I assumed it was the very same species I had encountered in northern Minnesota.

Although this is not snake country, I came upon a red-bellied form of the Eastern Garter Snake when a small individual crawled up my leg as I kayaked on Middle Menako Lake. Frogs and toads were heard, not seen; Spring Peepers sang at Second West Lake. Boreal Chorus Frogs vocalized in a wet bog.

Of course, mammals are the high-profile feature of this wilderness realm. I'd learned from my pretrip review of my mammal field guide that many species inhabit this patch of the North Woods. I had dreamed of wolves and Moose and caribou but discovered instead that American Black Bear is the commonplace large quadruped, on the move in the month of June. Aside from bear, the only common predator here is the Red Fox, one of which came into my Pipestone camp daily to scrounge for scraps. I often heard him rummaging around in the food box in the early morning while I was still in my tent. I also spotted tantalizing signs of wolves: paw tracks in wet sand, and droppings filled with the hair of their prey. I never heard them howl, however, and the only wolves I actually saw were those that visit the Musselwhite mine dump.

Other animals were surprisingly scarce. I saw Moose cows on only two occasions, in spite of the abundance of waterways and wetlands and bogs, and although I'd expected to spot Moose while kayaking, I was disappointed. I found a porcupine kill, possibly by a Fisher, but never actually saw a live Porcupine or Fisher. Nor did I see Woodland Caribou or Wolverine, although a local First Nations resident spoke of these species wandering through from time to time.

Smaller mammals were uncommon, too. I saw a single Snowshoe Hare, and a Woodchuck in a grassy clearing. Several times Least Chipmunks came into my Menako Lakes camp, and I encountered Red Squirrels only three times in sixteen days. Among the other mammals I *never* saw in my stay in the North Woods were White-tailed Deer, American Mink, Least Weasel, River Otter, American Marten, Striped Skunk, Coyote, Beaver, or any bat species. The camera trap I set up on several game trails caught images of only two vertebrates—the Gray Jay and me. (We'll ruminate on the reasons for the region's relative paucity of mammals in a later section addressing the possible impact of trapping for the fur trade.)

The local flora provided more satisfaction. Marsh Marigold was blooming. Various roadside shrubs were in flower: Pin Cherry, shadbush, and Labrador Tea. A dwarf blueberry showed its pendant bell-like white flowers, with their narrow pink fringe. Pink Ladyslipper orchids were rampant in some clearings: I came across a cluster of fifty-six in one spot. Dandelions and wild strawberries were in flower, as was the very small False Lily of the Valley. At Badesdawa Lake, plenty of Wild Sarsaparilla grew as ground cover at the forest edge, and *Clintonia*, Bunchberry, and American Mountain-ash were in high flower. It *was* spring, after all.

MATE ACQUISITION AND NESTING

Cooking dinner one evening at camp, I listened to the sweet *wita witeeyu* song of a Magnolia Warbler as I cut onions for hash. I located its nest—the only songbird nest I found during my entire journey—in a small spruce. The Magnolia is one of the more common boreal warblers, but here, where the territorial warblers evidently were scattered far and wide in low densities, I'd heard only one or two a day. The Magnolia is a commonplace migrant along the East Coast but remains popular because of the male's striking coloration—it is trimly patterned in yellow, black, white, and green, with a black eye-stripe, white eyebrow, large white wing-patch, and black flank streaks. It breeds in the Appalachians, New England, and much of boreal Canada. A confiding bird of the understory, it allows birders to goggle at its obvious beauty.

Earlier, this chapter discussed the mystery of how northbound migrant songbirds locate their natal territories in the northlands. Now let's turn to the questions of mate acquisition and nesting—activities constituting the crescendo of the life cycle of these northward-traveling songbirds—using the Magnolia Warbler as an example. How does a male Magnolia Warbler benefit by returning to his natal territory? The best explanation is familiarity. This is the place his parents successfully raised him, and this is the first patch of woods where he learned to forage for himself. The fact that he remains alive indicates that he got a good start in life, and that fact in turn indicates his natal territory was productive. By returning there, he has a head start on his nesting challenge. If his parents are back in place on his natal territory, he usually can take up a territory nearby. But because of short life spans, the male Magnolia's parents might have died, leaving his natal territory open for occupation. For an older male that has already successfully bred and is now returning north, it pays for him to return to the place where he earlier raised offspring—it is a place of proven quality, where he already knows the ins and outs.

The male sings most frequently and loudly in the early-morning hours to establish himself on his territory. At other times in the day, he sings to attract a mate. Presumably the songs he sings for the two tasks are distinct. Soon after his arrival on territory, a male will sing hundreds or even thousands of times a day, from dawn until dusk. (The literature reports on one male Red-eyed Vireo that sang more than twenty thousand times in a single day.)

The Magnolia female arrives about ten days after the male and goes about selecting her mate based on territory quality and male song and plumage brightness. Once paired, the two birds mate many times and search out a nest site. Both sexes contribute to nest-building, though the female does the most. The nest, situated about five feet up in a small conifer, is a flimsy cup of coarse grasses set atop a platform of twigs and lined with horsehair fungus. The female lays four eggs in four days and begins brooding the clutch after the fourth egg is laid; she alone will incubate. Once the eggs hatch, after about twelve days of incubation, both parents feed their nestlings tiny mashed-up insects, with the male delivering more food to the nest than the female. The young depart the nest after about nine days and begin learning to forage independently. The parents provision the fledged nestlings for several weeks, until they can forage on their own. If the initial nesting attempt is successful, the Magnolia Warbler pair does not re-nest that summer. If the first nest fails, the pair may attempt another nest. For the mated pair, getting several young to independence is their goal for the year.

What happens on the breeding ground, of course, impacts the number of birds that arrive on the wintering ground. Studies by John Faaborg and his students in Puerto Rico indicate that drought on the breeding ground leads to lower numbers on the wintering ground. Climate change is likely to influence rainfall patterns in North America, and this will certainly influence breeding productivity from place to place—the wetter sites will be more productive. What the overall net effect will be is uncertain.

During my sixteen days up north, I shared the campgrounds with three other groups, all fishing for Walleye. At Pipestone, a family fishing group from rural Michigan was camped when I arrived and settled into an adjacent campsite. The group was on its annual summer fishing adventure. A group of nine, spanning three generations, they were very friendly, and the morning I met them they offered me a luxurious hot breakfast of link sausages, a cheese omelet, and blueberry pancakes with maple syrup that they had manufactured themselves back home. The elder of the Michigan group mentioned seeing a wolf in the campground just before dawn that morning. As we chatted, I found out that they hailed from Fairview, in the northern section of the Lower Peninsula, deep in Kirtland's Warbler country. I told them I had plans to visit their local environs in a couple of weeks to see a research study on the Kirtland's Warbler, one of the last species on my quest list of wood warblers.

THE MISHKEEGOGAMANG AND HUDSON'S BAY COMPANY

I had planned to pay my respects to the chief of the Mishkeegogamang Band, but the death of an elder during my time in Pickle Lake prevented that, and I never spoke to a single member of the band while in the North Woods; neither did anyone from the community visit me at my camp. First Nations people in big pickup trucks zoomed past as I biked along the gravel highway, and the many hunting camps I passed stood unoccupied at this time of year. These camps—tiny, patched-together single-room cabins set in roadside clearings—serve as autumn and winter bases for fishing, hunting, and trapping. The forest, with its fauna and flora, is the wilderness supermarket of the Mishkeegogamang.

I did learn a bit about First Nations worldviews from discussions with an Ojibwe-Cree couple, Martin Kanate and his wife. From the North Caribou Lake First Nation, they keep a mobile home right beside the Nord Road at the Pipestone River bridge because Martin has the district government contract to grade that section of the road. We met when the couple came down to the Pipestone campsite to get fresh drinking water from the river. Martin was very friendly and talkative; his wife remained in their pickup, listening to our conversation through the open window. Both were big Moose hunters, and at one point Martin's wife piped up, "The only good Moose is a dead Moose." She assured me that she shot every Moose she saw. This shocked my environmentalist ears at the time, but Moose and fish are the staples of life for remote First Nations communities here, which have minimal access to grocery stores. I asked Martin lots of questions about local wildlife, especially wolves. He assured me that he killed wolves on sight, saying the animals are wasteful and often kill just for pleasure, a statement best understood as coming from a subsistence hunter who believes that these major predators take food from First Nations tables.

In later conversations with Kanate when I encountered him on the Nord Road, I came to realize that it was probably impossible for me to see the North Woods through his eyes. Our experiences and viewpoints were different: for one, I was not an indigenous North American, nor had my relatives been taken from their birth families and forced to attend a government boarding school far from home for their childhood years, among other cruelties. I have since read Marj Heinrichs and Dianne Hiebert's book *Mishkeegogamang*, which seeks to capture the essence of this people, and I have done additional research on the history of First Nations communities inhabiting the North Woods. I have learned that, as with most human societies, local history is very messy and obscured by a lack of details and by the fact that the history of Canada has been written largely by white

Canadians, with only minimal consideration of the travails of the First Nations peoples. History, after all, is about power.

For many decades, the power over the wilderness of northern Ontario was held by a venerable trading enterprise: Hudson's Bay Company. Receiving a royal charter from Great Britain in 1670, the company promulgated the harvest of fur pelts of an array of forest-dwelling mammals in a vast tract of north-central Canada then called Rupert's Land. Rupert's Land encompassed about a quarter of Canada's non-icebound territory—larger than today's Ontario. For many decades, Hudson's Bay Company was the de facto government in these undeveloped lands. The citizenry of Rupert's Land consisted of First Nations groups living and trapping there and an assortment of Anglo fur trappers and traders. During the fall and winter of each year, First Nations and European men trapped and prepared pelts, then traveled by canoe and on foot to Hudson's Bay's trading posts to sell the pelts. In exchange, they typically received popular trade goods such as knives, kettles, beads, needles, and Hudson's Bay point blankets. Sustained primarily by the trapping of beaver pelts to satisfy the European demand for felt hats, the intensely competitive trade opened the wilds of central Canada to exploration and settlement; financed missionary work; established social, economic, and colonial relationships between Europeans and First Nations bands; and played a formative role in the creation of the nation of Canada, which achieved independence only in 1867. During this long period, the fur trade was a huge influence on indigenous inhabitants.

Fur trapping remains an important occupation in the North Woods today. A 2013 article in Canada's *National Post* outlines the details: North American Beaver, which drove the initial trade, remains the second most valuable pelt in the trade, generating more than two million dollars a year. The article provides a list of the number of animals trapped and killed in Canada in 2010 to supply the international fur trade: 265,000 Muskrat, 140,000 North American Beaver, 92,000 American Marten, 47,000 Coyote, 28,000 Ermine, 16,000 Fisher, 9,000 River Otter, 7,500 Canada Lynx, 2,900 Bobcat, 2,900 Timber Wolf, 2,000 American Black Bear, 260 Polar Bear, and 8 Grizzly Bear.

These rather chilling statistics lead to several conclusions. First, virtually every watershed in the North Woods has suffered annual commercial harvests of wildlife for more than two centuries. In addition, the incidental by-catch of birds and nontarget mammals by trapping is probably double or triple the totals of the harvest numbers cited above, because the traps are not species-specific. Over the decades, trappers in the field also killed additional millions of edible mammals and birds for subsistence consumption.

The most significant impact of trapping is that the strong economic incentives of commercial wildlife harvest drove wholesale geographic movements of populations of Anglo and First Nations peoples across the landscape. As an example, the Mishkeegogamang (a band of the Ojibway nation) are not the original tribal inhabitants of the land north of Pickle Lake, where I surveyed birds. The area was historically Cree land. Apparently, driven by the economics of the fur trade, Ojibway bands from the Great Lakes region moved northward to trap wildlife, driving the Cree into more northerly territories.

Over the past century, the lives of the First Nations bands here probably have been more influenced by Christian missionaries, Hudson's Bay Company trading demands, mining interests, and Canadian government interventions than by any cultural imperatives of traditional First Nations ways of life. Little of their lives, as I observed them up north, seemed ancient or traditional, but instead seemed an unfortunate by-product of the economics of western business interests. It was a contrast to my decades of fieldwork experience in rural Papua New Guinea, where forest-dwelling indigenous peoples remain the owners and masters of their lands and resources. In spite of decades of Australian colonialism, local cultures, traditions, and languages survive in Papua New Guinea, and the spirit of the people is strong and confident.

The existential angst I sensed among First Nations people at Pickle Lake may be the end result of many decades of buffeting by outside forces. Poverty, loss of cultural traditions, loss of tenure over the land, and harmful government interventions have impacted indigenous groups inhabiting the wilds of central Canada. The human/natural ecosystem that I experienced in the North Woods may be the result of a struggling indigenous population inhabiting an unproductive landscape, which has driven the substantial overharvest of the wildlife resources that underpin the local economy. Why did I see so few mammals? Probably because the Jack Pine barrens have a low carrying capacity, but also probably because chronic year-round harvest of meat and pelts keeps mammals scarce. The wild mammal population in suburban Bethesda, Maryland, where I live, is likely considerably larger than the one in the habitat I visited in the distant woods of northern Ontario.

I read *The Maine Woods* while in northern Ontario. Published in 1864, Henry David Thoreau's narratives from his back-country travels in Maine address four general themes that resonate in today's rural Ontario: (1) the terrible social conditions facing indigenous people living in an Anglo world; (2) the systematic overharvest of timber by rapacious timber companies; (3) the overhunting of Moose by resident indigenous people; and, of course, (4) the ever-annoying mosquitoes and black flies. Today, timber harvest in Ontario continues to have broad-scale impacts, although it is now substantially better managed, but it is astounding that the problems that concerned Thoreau in the mid-1800s in the boreal woods of New England are still relevant in a similar wilderness habitat in Canada today. Have we as a civilization progressed so little over those 150-plus years?

THE WILDERNESS, THE AMERICAN BLACK BEAR, AND THE MOOSE

When planning field adventures, city-based naturalists such as myself love to imagine wildlife-rich "wilderness"—some place far away and different, empty of humans and possessing fearful wild creatures. This myth is part of our historical American frontier legacy. We now know that such fantastic wilderness places never really did exist, that humankind long has occupied much of the landscape, and that people long have been harvesting most things of economic value from virtually every quadrant of the earth's landscape. In spite of these rude facts, we dreamers manage to convince ourselves that such imagined wildernesses do exist. I had fallen into this trap with regard to north-ern Ontario: I had driven to the end of the road—akin to the "end of the earth"—with mildly delusional visions of great conifer forests bursting with big, scary mammals and trees full of colorful migrant songbirds. What I found was a fire-scarred pine barren inhabited by an impoverished and underserved indigenous population who were living by their wits and largely forgotten by the rest of the world. This underdeveloped land was built upon a glacier-scoured shield of granite, with sand and gravel for soil. Because of past and present impacts of recurrent fire, an array of gold-mining operations, and the centuries-long fur trade, the land was scarred and bruised.

And yet it still holds treasures for visiting fishermen and wilder-ness canoers, who found the lakes and rivers perfect for their needs. And for the adventurous birder, it holds avian riches in small dol-lops, hidden in spruce bogs here and there, requiring considerable effort to find. I had to work hard to find my Connecticut Warbler, my Great Gray Owl, my Spruce Grouse, and my American Three-toed Woodpecker. And I had had to pay my dues. On warm and humid days, black flies and moose flies seemed to boil up out of the boglands—those were head-net days. Was I disappointed by what I dis-covered? Like a person who has had a glass of cold water thrown in his face at an unexpected juncture, I was shocked and brought up short, but the experience was bracing and memorable. I got what I deserved.

The wilderness I imagined was in my own mind, and I was able to conjure up a wilderness experience that fit my own psychic need. In fact the wilderness *was* that Connecticut Warbler in a tall row of Tamaracks. It was that Great Gray Owl in a stand of spruces. My wilderness was the ever-present concern that I might turn a corner and encounter a five-hundred-pound American Black Bear a few steps away (that's why I carried a canister of bear spray on my hip every time I went out). There were times in the field when I felt the presence of something wild, and the little hairs on the back of my neck stood up. I would then stand still, listen intently, and try to search out what it was. Perhaps this was the wilderness speaking to me, telling me to be humble, to be cautious. The American Black Bear is the king of the North Woods where I camped, because it is treated with respect by the First Nations people who live here. This is the ancestral home of the American Black Bear, and one can feel it. I saw eight American Black Bears during my stay, and I am sure at least eight more American Black Bears saw me during that time.

The Moose, on the other hand, is the local meat market. They are harvested to sustain the Mishkeegogamang, and thus I saw the animals infrequently. The Moose were doing their job, hiding in a thicket, waiting passively for that local hunter to collect them and to convert every bit of heart, liver, meat, sinew, and bone into something useful.

There is satisfaction in following the passage of songbird migrants from the coast of east Texas northward up the Mississippi and then finding them on their breeding grounds in northern Ontario. I found, among others, Tennessee Warbler, Magnolia Warbler, Swainson's Thrush, Ovenbird, Red-eyed and Philadelphia Vireo. But there was disappointment in seeing the North Woods so impoverished. Why did the forests here support so few birds?

One explanation is that this situation is a product of disparate geographies: winter versus summer. Subtropical and tropical habitat available to the wintering Neotropical migrant birds is rather circumscribed (mainly ranging from Colombia north to Guatemala and southern Mexico, plus the Caribbean islands). By contrast, the summer habitat available to breeding songbirds is massive—from Alaska and the Yukon southeast to New Brunswick, New England, and the Appalachians. Presumably the small geographic extent of the tropical and subtropical wintering habitat can sustain only so many wintering Neotropical migrants. When, at the end of northern winter, the surviving pool of wintering birds moves back north, it is greatly diluted by the vast expanses of the boreal forest—hence the emptiness I encountered. This great disparity in available winter versus summer habitat might explain the paucity of migrant songbirds in the boreal forest. My position in the middle of a Jack Pine barren simply added to the impoverishment.

On June 13, the voice of a Swainson's Thrush in a young White Spruce just behind my tent woke me at 3:45 a.m. The air temperature was near freezing, so I put on a woolen watch cap and long underwear before heading out into the chill morning air in search of something wild. Out bicycle-birding, my fingers tingled with cold in spite of my work gloves—that's the weather one comes to expect in the far north. I finally encountered a flock of Pine Siskins, a new bird for the list. At 2 p.m., I sat on a rock and watched hundreds of cumulus clouds scud by through the huge expanse of deep blue sky across the lake. It reminded me of a summer day in the Adirondacks in 1964. In that summer I had my first taste of the deep woods. I climbed my first mountain, saw my first American Black Bear, and caught my first Smallmouth Bass.

After dinner I read *Annals of the Former World*, a big, fat John McPhee book about the geology of North America. It started to get dark around 10 p.m. A full moon was rising, and the Swainson's Thrush was singing, once again, behind my tent. Here, in my little circumscribed world, I remained out of touch with the world at large. I was living full, rich days and felt no need for a newspaper, radio, or TV. The mosquitoes were starting to swarm outside the tent, their vast numbers producing an audible hum. I was woken briefly at 11:30 p.m. by a vocalizing pair of Common Loons as they flew overhead. Then I heard the drumming of the Ruffed Grouse—he must have been moonstruck.

In my sojourn up north, I had made observations of eight additional breeding wood warblers species on their breeding habitat, bringing my total to thirty-two. I now had five more to track down on the remainder of my quest.

EIGHT

Great Lakes Country

Late June to Early July 2015

The longest day of sunlight . . . comes at the beginning of Summer

rather than in its midst. In consequence, all Summer long we are

inclining towards Summer's end instead of building to a climax and

then tapering off.

—HAL BORLAND, *Sundial of the Seasons*

On June 21, I rise at 4 a.m., break camp, and head down the Nord Road to Sandbar Lake Provincial Park, not far from the town of Ignace. I am heading back into civilization after sixteen days in the far north. Sandbar Lake is 225 miles south of Badesdawa Lake, and today it is the summer solstice. I am on my way, by a circuitous route, to a mountaintop in the Adirondacks, where I will hunt for my last breeding wood warblers and the last vestiges of spring.

I looked forward to shifting south into biologically richer territory and away from the scourge of the fire-prone Jack Pine. The drive from Pickle Lake to Sandbar Lake is famous as a route to see big game, and my haul was fair: two cow Moose, a Red Fox, a Snowshoe Hare, a tiny unidentified rodent, and two hen Ruffed Grouse, each with a batch of fuzzy peepers. I crossed a high ridge with a wide vista of forest near the Sturgeon River, passed through the tiny hamlet of Silver Dollar, and came upon roadside scenes of brutal clear-fell logging just south of Savant Lake. Here I saw, for the first time, big stands of roadside Bracken fern (I had seen no ferns up north). About twenty miles north of Ignace, tall pines stood out as welcoming sentinels above the forest canopy. I was now back in White Pine country, more familiar territory.

Sandbar Lake is a tidy but unremarkable provincial park. SUV-sized glacial erratics stand in the campground—a reminder of the recent history of glaciation here. The area has a southerly aspect, with abundant ferns and a forest of Paper Birch, Quaking Aspen, Balsam Fir, White Pine, and White Spruce. The Pileated Woodpecker was common here, a species absent farther north. At 5 p.m., many birds sang, and in the evening, I sat writing field notes at my picnic table, which was mosquito-free (when I'd left my car door open while departing my more northerly camp this morning, the car filled with the voracious pests). As the night sky darkened, I heard a Swainson's Thrush. It had sung for me in evenings from Arkansas to Ontario.

OPPOSITE: Blue-headed Vireo

The next morning I walked the campground circuit, which was birdy. At an empty campsite just down the loop from my own, the high, weak, slurred *si-syu . . . si-syu . . . si-syu* of a male Bay-breasted Warbler—quest bird number 33—sounded. This elusive creature, which I'd seen last in Texas as a passage migrant, and for which I'd searched spruce stands high and low in the far north, was here at a drive-in campground. I spent several hours the following two mornings photographing the male Bay-breast doing his territorial thing. A close relative of the Blackpoll Warbler, the Bay-breast breeds in boreal conifer forests in the northernmost sectors of the U.S. border states between Minnesota and Maine, as well as from eastern to northwestern Canada. After wintering in northern South America and the Caribbean, the Bay-breast migrates north through the eastern and central United States. It favors breeding habitat that is experiencing a spruce budworm outbreak (a phenomenon discussed later in this chapter). The male is strangely handsome, with a deep chestnut crown, throat, and flanks, a black face mask, dark-mottled upper parts, and a whitish breast. Along with the Cape May Warbler, it is a favorite of many birders. This particular male on territory flitted about, foraged, and sang incessantly from the campsite, and it brought me back onto the track of my quest: only four species of wood warblers to go!

At Sandbar Lake, I also added a new quadruped to the trip list: Masked Shrew, identified by its brown pelage and long, brown fur-covered tail. At five grams, this was the smallest mammal I had ever encountered. Back at my campsite, I made another little friend—an Eastern Chipmunk. The little guy held my finger in its two paws to receive food and allowed me to pet him while he foraged placidly on the peanuts I handed him. Clearly, he had had experience with humans and knew about people food prior to my arrival. Such a confiding Eastern Chipmunk is hard to resist.

I biked along a narrow path through spruce woods by a wetland. Blue Flag iris bloomed in the beaver meadow. Mountain Maples flowered throughout the open woodlands, showing off their erect spikes of pale-yellow florets. Along this path, I recorded Wood Duck,

North on the Wing

Bay-breasted Warbler

Black-backed and Hairy Woodpeckers, Swainson's and Hermit Thrushes, Least Flycatcher, Common Raven, American Redstart, Canada, Ovenbird, Red-eyed, and Blue-headed Vireos, Swamp and Song Sparrows, and Nashville, Blackburnian, and Tennessee Warblers. I had never racked up such a long list so quickly in the far north—it was a luxury to return to such bird-rich woods. Then I biked a few miles up the main road, only to be shocked by the scale and destructiveness of logging here—even right by the roadside. I hurried back to the protected forests of Sandbar Lake.

RESOURCE EXTRACTION AND THE FUTURE OF CANADA'S BOREAL FORESTS

The broad-scale logging in the Sandbar Lake area got me thinking about Canada's boreal forests. Before my trip, I had heard about threats to this massive ecosystem, and I'd also learned of conservation initiatives to address these threats. I had read about the Canadian Boreal Conservation Framework, a partnership established in 2003 that links a dozen organizations and government departments to conserve boreal forest. Scientists from around the world, dozens of major companies, Canadian First Nations people, and Canadian and international environmental organizations now work with the program to better protect Canada's 1.2-billion-acre boreal forest ecosystem. The goal is to preserve at least half of this forest in a series of connected reserves, which will allow the vast ecosystem to function as it should. Preserved forest will provide habitat for migrating caribou and for production of the ecosystem services that underpin the health of the hemisphere.

The challenge, of course, is to balance adequately economic development with smart conservation action. The task is made complex by the disparate viewpoints of stakeholders in this discussion; common ground must be found among miners, loggers, pulp producers, First Nations groups, and the conservation community. Yet there has been progress: in 2010, Ontario passed the Far North Act, which set targets for strict protection and for sustainable development of 110 million acres of boreal forest. It was the first provincial-scale law in Canada to balance development with forest conservation. This bill, in addition, provides for community-based land-use planning by Ontario's First Nations.

Also in 2010, Canada's forest industry and several leading environmental groups signed the Canadian Boreal Forest Agreement, in which timber companies agreed to temporarily suspend operations in roughly 71 million acres of forest, representing a significant portion of habitat for Canada's threatened Woodland Caribou population. The agreement improves sustainability practices on at least 108 million acres. In all, more than 350 million acres of boreal forest in Canada are either strictly protected or have been committed for protection. Furthermore, governments have pledged to sustainably develop more than 350 million additional acres, using ecosystem-based resource management practices and state-of-the-art stewardship practices. This is heartening news, although the devil is in the details, and the negotiations will continue for decades to come.

A nagging but unanswered question is how clear-fell logging of Canada's boreal forests has impacted the Neotropical migrant songbirds that breed here. I suspect these landscape-wide timber extraction activities benefit some early successional species but harm those that prefer undisturbed closed forest and old-growth conifer forest. Field studies have shown that permanent resident species such as the American Three-toed and Black-backed Woodpeckers are harmed by short-rotation clear-felling that is followed by intensive silvicultural activities to enhance yield. These two special woodpeckers benefit instead from the presence of mature conifers and the effects of periodic wildfire.

THE NORTH SHORE

On June 23, I say goodbye to my adorable chipmunk friend and travel 278 miles from Sandbar Lake to Rainbow Falls Provincial Park. The route takes me to Ignace, Thunder Bay, Nipigon, Rossport, and then Rainbow Falls, which is near Terrace Bay. I pass an abundance of mature forested habitat but see very little wildlife—only a single Red Fox. I am now on the Trans-Canada Highway, heading east toward Ottawa. Along a stretch of highway are stands of Balsam Fir killed by a spruce budworm infestation. I cross from the catchment of Hudson Bay into the catchment of the Saint Lawrence. East of Thunder Bay, northeast toward Nipigon, Balsam Fir takes the place of White Spruce as the dominant conifer. I enter substantial and forested rolling hills, and come upon even more massive hills, buttes, and basalt cliffs on the north shore of Lake Superior. At one overlook, a spectacular vista across the north shore reveals a giant table mountain with fjordlike inlets. Toward Rossport, the north shore is very rugged, with high ridges and bays and islands—all forested and looking a bit like Nova Scotia. The giant lake to the south seems positively oceanic. Late in the day I arrive at Rainbow Falls, with its hilly landscape of Red and White Pines. The evening is cold and damp, and the black flies swarm.

The north side of Lake Superior is largely unfamiliar to those living in the United States, who mainly know the *southern* shore of Lake Superior—including the Upper Peninsula of Michigan and a bit of northernmost Wisconsin—as a faraway place. Cross the giant lake due north from Marquette, Michigan, and you arrive at Rainbow Falls, where I was camped. Although I'd returned from the distant Ontario northlands, I was still well off the beaten track: the nearest city was Thunder Bay, three hours to the southwest. This is the land of long distances, few roads, and big lakes.

Next morning, I birded around my campsite, finding Black-throated Green Warbler, Ovenbird, American Redstart, Least Fly-catcher, Red-breasted Nuthatch, and Red-eyed Vireo. After breakfast I hiked to the summit of the Back Forty Overlook, which provides spectacular views to Superior's rocky north shore and a string of rugged islands to the south and southwest: Copper Island, Wilson Island, Vein Island, and Simpson Island. The clouds blew off, blue sky took hold, and a crisp, springlike day emerged. On the rocky overlook were Nashville and Magnolia Warbler, Black-capped Chickadee, Common Raven, Swainson's Thrush, flowering Bunchberry, *Clintonia*, Pink Ladyslipper, False Lily of the Valley, and Crowberry.

In late morning, I headed southeastward to Pukaskwa (pro-nounced "PUCK-a-saw") National Park, on the northeastern shore of Lake Superior. Leaving the Trans-Canada Highway at Heron Bay, I drove south to the Ojibway First Nations community at Pic River and crossed the gravel-strewn White River to the park entrance. Right beside the road sat a rock the size of a house—presumably a supersized glacial erratic. Here spring was still in full flower: I saw a Mourning Cloak butterfly and a Canadian Tiger Swallowtail at the park entrance, as well as a Shadbush in flower, and at the visitor center, Pin Cherries blossomed brightly.

The day transformed itself once I got to my campsite on the shore of Lake Superior. A heavy fog lay on the water and edged into the shoreline forest. It was like being on the coast of eastern Maine: chilly and dank. I was camped in a low, dense conifer forest set on shield rock next to the cold lake. White Spruce, Balsam Fir, Northern White-cedar, birch, and aspen dominated the woods. This would be home for the next few days.

That afternoon, I took a long bike ride on the well-paved, flat roads, traveling north to Pic River. The local First Nations people living here are the Anishinaabe, a major group encompassing the Ojibwe, among others. First Nations people represent 4.3 percent of the Canadian population, versus the 2 percent that Native Americans compose of the population of the United States. As I biked around, my hands ached in the fog-chilled air. Later, back in the park, I walked to Halfway Lake and lay on a rock overlooking the water, just relaxing, which is something I had not done much of on this trip. It was peaceful, and the black flies and mosquitoes few. As I lay on a carpet of low vegetation, taking in the tranquility of it all, I heard the voices of two species, Myrtle Warbler and Common Grackle. Returning to my camp, I found a raven eating my loaf of bread. He flew off with the remnants as I approached. Over the weeks, my food stores had been raided by a Common Raven, a Raccoon, a Red Squirrel, a Red Fox, and an Eastern Chipmunk.

The fog broke in the evening, and the blue sky opened up to make the end of the day pleasant. I heard the voices of Ruby-crowned Kinglet, Common Raven, American Crow, Magnolia Warbler, Black-throated Green Warbler, American Redstart, and Hermit and Swainson's Thrushes. At 9:30 p.m., a Swainson's Thrush—my musical sentinel of the boreal realm—sang in the adjacent thicket. The temperature (44°F) and the thrush were a lovely combination for heading off to dreamland.

I rose to greet a windless, crystal-blue sky, with the temperature a bit above 40°F. Before breakfast, I biked the local roads in search of birdsong and walked the perimeter of Halfway Lake, producing good wood warblers: seven Nashville, seven American Redstart, two Mourning, a Northern Parula, a Myrtle, a Tennessee, a Magnolia, a Canada, a Black-throated Green, and a Common Yellowthroat. I had recorded ten species of warblers in a *couple hours* of morning birding. In sixteen days in northern Ontario, I had tallied thirteen species of warblers total (and never more than eight species in a single long day). I now had truly left the empty northlands behind.

I encountered a female Spruce Grouse with a brood of little fuzzy brown chicks. I saw two Snowshoe Hares, both dark-pelaged except

Myrtle Warbler

for their oversized white feet, and I managed to add two new birds
to the trip list: Red Crossbill and Bank Swallow. The crossbills flew
over in a tight flock, giving their *kip-kip* calls, and the Bank Swallows
swarmed in the air beside their nesting colony in a sandy cliff cut by
the White River. I found my first Northern Parula in Canada, near the
northern limit of the species' breeding range. Recall that this was the
very first forest-breeding warbler I'd encountered back on the coast of
Louisiana.

The hike around Halfway Lake this morning produced more
sublime solitude. I was serenaded by the lonesome song of the White-
throated Sparrow and a single vocal Spring Peeper. A hen American
Goldeneye made ripples on the glassy lake, and silently flew off when
it noticed me. An adult Common Loon came streaking high over the
lake on its way somewhere.

In the late morning, I hiked up onto the park's open, rocky
promontory, which overlooks the expanse of Lake Superior. On the flat
summit, colorful plastic Adirondack deck chairs beckoned me to sit
and gaze over the water. There was no hint of wind, no cloud in the
sky, and the temperature was in the sixties—no intimation of summer
heat or humidity here. Nearby, a Ruby-crowned Kinglet, perched atop

a spruce spire, sang his heart out in the full sun; this little mite is one of the rock stars of the boreal forest. A few minutes later, an adult Bald Eagle soared slowly overhead, trying to ignore several heckling gulls.

A NATURAL HISTORY OF THE GREAT LAKES

After two nights in Pukaskwa, I rise in the 37° F dawn and head due south to Luzerne, Michigan, where I have a date with Kirtland's Warbler. I trace the eastern flank of Lake Superior toward Sault Sainte Marie and the U.S. border. I drive from Heron Bay to White River on the Trans-Canada Highway through rolling, rocky, forested country that reminds me of the Adirondacks. A thick fog obscures the road at White River. Escaping the blanket of fog, I find more shield rock, hills, spruce, aspen, fir, and curvy roads for fun driving. Few cars pass on the road in this morning's pleasing desolation. Stark yet beguiling, the lonely road shows me evidence of glaciation everywhere. South of Wawa, the highway runs through the heart of Lake Superior Provincial Park, very rugged and mountainous and covered in boreal forest. The roadside birdlife, mainly ravens and crows, is sparse.

The five North American Great Lakes constitute the largest cluster of freshwater bodies on earth, holding 21 percent of the earth's surface fresh water (excluding the planet's ice caps). The lakes began to form during the retreat of the Wisconsin glaciation about fourteen thousand years ago. The great ice sheets carved the lake basins, which then filled with glacial meltwater. It is interesting to note that in spite of their current environmental and economic importance, the Great Lakes are not old and are in no way permanent. All lakes of the world are ephemeral. As John McPhee has written in his magisterial *Annals of the Former World*: "Lakes fill in, drain themselves, or just evaporate and disappear. They don't last." It is difficult to think of the upper Midwest without those vast, sealike bodies of water, but that may come to pass in a geological blink of the eye.

Still, on the human time scale, the lakes are a big deal, and I was in awe wherever I came upon them. The lakes drain into the

Atlantic through the Saint Lawrence River. The fresh water flows from lake to lake in a series of steps, from the highest—Superior—to the lowest—Ontario—until the water flows from that lake into the Saint Lawrence. But the details of this system are complicated. The waters of Superior flow east through the rapids of Saint Mary's River and the Soo Locks to the North Channel of Lake Huron. Just south of Sault Sainte Marie, Lake Michigan flows east through the Straits of Mackinac, on the south side of the Upper Peninsula, into Lake Huron. At the southern tip of Lake Huron, the Saint Clair River drains south into Lake Saint Clair, and then Lake Saint Clair drains south through the Detroit River into the western end of Lake Erie. The eastern end of Lake Erie drains north through the Niagara River, down Niagara Falls, and into Lake Ontario. Finally, the northeastern terminus of Lake Ontario funnels into the Saint Lawrence, which flows eastward and northward into the Gulf of Saint Lawrence and finally the Atlantic.

Lake Superior (known as Gitchi Gami in the Ojibwe language) is the superstar among these giant bodies of fresh water—the size of Austria, it is 1,330 feet deep and holds 12,100 cubic kilometers of water. In its depths, Superior is a uniform 39°F. Because of its geographic position, the lake generates sizable waves, sometimes reaching 30 feet.

KIRTLAND'S WARBLER

The Smithsonian's Pete Marra had counselled me to visit his project studying Kirtland's Warbler in Michigan. This was the appropriate moment to do that, given my relative proximity to his field site in Luzerne, Michigan, due south of the eastern shore of Lake Superior—and four hundred miles south of Pukaskwa National Park. I crossed the international border at Sault Sainte Marie, my first encounter with a real city since my brief stop in Duluth about a month before. I suffered a bit of culture shock from the early afternoon's traffic mayhem. Once in the United States, I hopped onto Interstate 75 and tore across the Upper Peninsula to the big bridge over the Straits of Mackinac, which separate Lake Michigan from Lake Huron.

The high-speed vistas from I-75 showed lovely expanses of green meadows filled with glowing beds of Yellow Rocket, as I had seen weeks earlier in southern Missouri. Continuing south on I-75, I glimpsed a pair of Sandhill Cranes just off the highway near Gaylord. Exiting the highway at Grayling, I took Route 72 east to Luzerne, in the north-central Lower Peninsula.

This is the thick of Kirtland's Warbler country. I was here to visit Nathan Cooper, a Smithsonian postdoctoral researcher working on Kirtland's in partnership with the Upper Mississippi and Great Lakes Joint Venture, the U.S. Fish and Wildlife Service, the U.S. Forest Service, and the Michigan Department of Natural Resources. Cooper and his team of five research assistants were focused on the nesting success of the warbler, the impact of Brown-headed Cowbird nest parasitism, and the movement of the warblers from their breeding to their wintering habitats, as informed by tiny geolocator devices affixed to the birds. The team spent its field season working at several breeding sites, hunting for nests, and netting birds to recover geolocators attached to the birds the previous summer. Once they found the nests, the team followed them through the hatching and fledging cycle to check on productivity and to determine the impact of nest parasitism by the cowbird in relation to distance from active cowbird removal operations. Cowbirds have been locally trapped and removed annually since 1972 to foster higher nesting productivity by the endangered warbler.

Cooper and his team shared a rented house in rural Luzerne, and I put my tent under a tree in their front yard. I then took the research group out for dinner at a nearby watering hole named Ma Deeters. We ate big burgers and drank cold beer in a big, noisy, wood-paneled room and talked research and birds, exactly what ornithologists like to do on a Friday evening in June.

While camped in Ontario, I had searched eBird online for late-spring sightings of Cape May Warblers—one species that had, to date, eluded me. I found one recent record of the species from Luzerne. Remarkably, that eBird record had been uploaded by one of Cooper's assistants, Ethan Gyllenhaal. Gyllenhaal had located several singing male Cape Mays on territory in conifers at Cooper's rental house as well as at the nearby Luzerne Boardwalk. The Cape May Warbler was a species that had remained just out of reach throughout my trip, yet *here* it was, singing in Cooper's front yard, at the very southern limit of its breeding range. I had journeyed here to spend time with the very rare and restricted Kirtland's Warbler but now found that the ever-elusive Cape May was thrown in as an unexpected bonus. In my tent by 10 p.m., I drowsed to the sound of a drumming Ruffed Grouse and dreamed of the two rare warblers that awaited me in nearby habitat.

On the next morning, Saturday, June 27, the field crew planned to head out as usual, because they worked seven days a week. I awoke at 4:45 a.m. to a singing Whip-poor-will and, after a quick breakfast, headed out with Cooper and David Bryden (an assistant from New Zealand) to a nearby Kirtland's study site. Along the way, Bryden told me he had visited my most recent field project in New Guinea. Ornithology is a small world, and field assistants and volunteers travel the world in search of interesting research opportunities.

Cooper told me that the Kirtland's Warbler population now stood at two thousand pairs—up from a low of just two hundred pairs in the late 1980s. The species' breeding habitat is essentially confined to a ninety-mile-diameter polygon of sandy outwash plain in the northern portion of the Lower Peninsula of Michigan, centered on the community of Grayling. Here I was back in postglacial Jack Pine habitat. The warblers set up territory only in very young Jack Pine monocultures, where the trees stand five to twelve feet tall. The birds nest on the ground in the shade of a pine, hidden in low, thick mats of blueberry. The males arrive on territory in early May, and the first nests are completed in late May. Nestlings typically leave the nest in late June.

This morning, we encountered at least six warblers, but because of the lateness of the season, the females sat quietly on the young and the males did not vocalize much, making them more difficult to track down. I followed Cooper and Bryden around plot DNR-2 as they checked the status of nests that had been located earlier in the season.

This patch of habitat, DNR-2, had been created solely and specifically for Kirtland's Warblers by conservation teams from the state, and its trees were seven years old.

Each Kirtland's nesting patch is manufactured by conservationists. First, the area is clear-felled, the trees and tree waste are removed, and the bare field is disked; pine seedlings that are grown in a nursery are hand-planted in rows in the cleared area, with an intertree distance of about nine feet. Herbicide is applied regularly during the growing season to keep down competing vegetation and to allow the pines to prosper. Once the trees reach a height of five feet, territorial male warblers begin to settle in the pine stand. At this stage, conservationists set and bait large wire traps in small clearings among the pines to capture and remove all visiting Brown-headed Cowbirds. Each site is closed to all visitors to allow the warblers to breed in peace. As of 2015, the state had twenty-three Kirtland's Warbler management areas, totaling 127,000 acres. Because each site outgrows its usefulness to the species in about seven years, the agency that owns the land needs to create new sites every year to make up for those that have outgrown the warblers. An additional several thousand new acres of breeding habitat are created each year.

Kirtland's Warbler is thus, like the Red-cockaded Woodpecker, a *conservation-dependent* species—one whose survival on earth depends on active annual intervention by humankind. For reasons unknown to us, the millions of acres of fire-prone Jack Pine monoculture that I found in northern Ontario are ignored by this rare warbler. There are plenty of fresh tracts of Jack Pine in Canada, but they remain empty of Kirtland's.

Kirtland's Warbler is also a poster child for the need for *full life-cycle conservation.* The term describes the need to take conservation action in all sectors of the species' year-round range to ensure its survival. Thus conservation scientists today are working to provide suitable protected habitat for Kirtland's Warbler in its restricted wintering range in the Bahamas, in its stopover sites in the southeastern United States, and, of course, on its breeding habitat in Michigan. In fact, full life-cycle conservation is a practice that would benefit scores of migratory songbirds, not just Kirtland's Warbler.

The Kirtland's breeding habitat is a bit off-putting to a first-time visitor. It has the look of a young Christmas tree farm, with trees planted in a geometric grid at a density of about 1,100 trees per acre (allowing for the presence of several openings within these plantings), something of a monotonous affront to those who love the randomness of nature. But Kirtland's Warblers appreciate these monoculture plots, so they are what conservationists give the birds. I found few other birds using the manufactured habitat: Hermit Thrush, Blue Jay, Common Raven, and Field Sparrow. Clearly, this is no avian hotspot, just a specialized nesting habitat for a very specialized bird.

Cooper's study included work on the Brown-headed Cowbird, which tricks other species into raising its offspring. The female cowbird mates and then lays its fertilized eggs into the fresh nests of a variety of songbirds, include Kirtland's Warbler. This trickery substantially reduces the nesting success of the species that receives the egg donation from the cowbird. The doting and naive parents end up preferentially provisioning the fast-growing and aggressively demanding cowbird nestling. The combination of the shortage of suitable Kirtland's nesting habitat, plus the drag of cowbird nest parasitism, had long kept Kirtland's numbers perilously low. The government recovery team, tasked with bringing the warbler back from the brink of extinction, has perfected a cowbird-trapping procedure that has massively reduced the incidence of nest parasitism of Kirtland's Warbler. This spring, Cooper's team located 150 warbler nests, and only one had a cowbird egg in it. After decades of summer trapping and removal, cowbird populations are now so low in central Michigan that they no longer pose an active threat to warbler nesting.

THE CAPE MAY WARBLER AND THE SPRUCE BUDWORM

After the visit to DNR-2, I joined the team of assistants on a quick tour of the Luzerne Boardwalk, a local birding hotspot that is the site of a boreal cedar swamp situated within Huron National Forest.

Within minutes of our arrival at the head of the boardwalk, we found a male Cape May Warbler singing on territory. Marked with chestnut cheeks, a rich yellow breast with abundant black streaking, and a dark crown, the male was collecting spruce budworm pupae and taking them to a nest high in a conifer to feed a nestling. A Lincoln's Sparrow sang from an adjacent field, as did a Black-billed Cuckoo and Magnolia, Nashville, Pine, and Palm Warblers, as well as several Ovenbirds. Driving back to Cooper's house, we saw a Coyote crossing the road. White-tailed Deer abound here—and fawns are prime food for Coyotes.

On Sunday, a White-breasted Nuthatch awoke me by calling outside the tent in the 39°F air. I headed to the Luzerne Boardwalk on my own, where the Black-billed Cuckoo was calling like crazy—was the cuckoo, a caterpillar specialist, celebrating the budworm outbreak? Nashville Warbler and Ovenbird were in song, as well as Rose-breasted Grosbeak and Lincoln's Sparrow. I had come to the boardwalk, however, mainly to spend more time with the nesting Cape May Warblers. Once again a male was collecting spruce budworms and carrying them up to its hidden nest. The foraging male was very confiding—at one point he perched down on the sandy walking path, chasing a writhing budworm pupa on foot as I stood eight feet from him. To East Coast birders, finding this species in breeding plumage in spring is a major challenge. The males are strangely beautiful, with their chestnut cheek and big white wing-bar. Males in spring migration on the East Coast mainly lurk high in towering Norway Spruces, their thin and high-pitched song easily missed. Thus it is a treat to see one of these gorgeous birds at close range and below eye level.

The Cape May Warbler is a spruce budworm specialist, and its populations, as well as those of the Bay-breasted Warbler, rise and fall with the abundance of the budworms, which they both eat and feed to nestlings. Because spruce budworm infestations are cyclical, coming and going in the boreal spruce-fir forest, Cape May Warbler populations follow a similar pattern. A small, nondescript moth in the family Tortricidae, the spruce budworm lays its eggs on the needles of spruce and fir. The larvae burrow into the needles and feed on them. Budworm infestations can kill a mature Balsam Fir, though spruce typically survive the attack. A single budworm cycle can last as long as two decades, populations building over years to a peak followed by a crash. The Luzerne area's spruce and fir were suffering the early stages of a budworm outbreak; I saw big red-brown patches of budworm-invaded needles on some Balsam Firs.

Midmorning, I headed out with the team to a second Kirtland's Warbler breeding site to spend more time with this second marvelous warbler. Yes, its breeding habitat is unprepossessing, but the bird itself is easy to love. The male sings a loud and chattery song from atop one of the myriad low pines. It is a glory to see him throw back his head, open his beak, and let his voice soar. The species is remarkably unwary; both males and females seem entirely unfazed by the presence of human observers. Thus birdwatchers fortunate enough to gain permission to walk in the habitat can share some intimacy with the birds as they go about their daily lives, and that is special for such a rare species; one can stand among head-high pines while the adorable little birds move about at eye level. I followed a Kirtland's pair for about twenty minutes, making a spishing sound to attract them, and in most instances the female approached within a few feet of me. Their plumage is classic for a wood warbler: mainly gray above and pale yellow below, with dorsal and ventral black streaking, plus a broken white eye-ring, thin white wing-bars, and prominent white undertail coverts. The eye-ring, set against the dark face, is particularly fetching. In many ways this is the prototypical wood warbler, yet strangely it is rare. When will it discover the vast Jack Pine stands of northern Ontario?

At some point in their lives, many birders decide to make the pilgrimage to Michigan to see the Kirtland's. Hundreds visit every spring. The warbler recovery team has established a protocol for visitors: they meet in Grayling or Mio, caravan to a breeding site, and carefully tour in groups led by a local volunteer to experience the bird and learn about the whole process of conserving the species, which includes the story of the cowbird. This is all very sensible and prudent, but, that said, something about seeing the species under such hothouse conditions detracts from the joy and independence of the birding experience.

One can avoid such an enforced group experience by seeing a Kirtland's on migration. Magee Marsh, Ohio, on the south shore of Lake Erie, is a particularly good stopover site to spot the species on its journey. In 2014, a year before this trip, I'd pulled into the Magee Marsh parking lot at about 6 p.m., after driving nine hours from home. A clot of photographers were hanging out, their tripods in a cluster in the grass beside the parked cars. I sauntered over to ask what was up. A Kirtland's Warbler was hanging out on East Beach. By this time, most of the day's visitors had returned to their motels after a long day of birding, but I drove to the East Beach parking lot and to search for what I thought might be a needle in a haystack.

To my surprise, I did not even need to locate the bird. I needed only to locate the remnant clusters of birders—a semicircle of five people in the low beach vegetation—communing with the Kirtland's as the sun dipped low. One turned to me and silently pointed to a piece of driftwood in the sand where a male Kirtland's perched, tail wagging diagnostically. I spent more than an hour on the beach as the bird foraged on the sand and in the vegetation, never straying more than ten feet from where I'd first seen him. Before long, I was the only person left with the bird. Ignoring my presence, he was intent on hunting down a type of small fly, abundant on the beach at that time.

HARTWICK PINES

Midday on Sunday, I break down my tent and head north to resume my Canadian adventure. Cooper's crew advises me to stop at Hartwick Pines State Park, just north of Grayling, before crossing the border. The park features a forty-nine-acre tract of old-growth forest and an adjacent museum dedicated to the history of the logging of Michigan's great virgin forests.

The park's circuit walk through the old-growth forest was more than worth the price of admission. It was a stand of White Pine, Red Pine, Eastern Hemlock, Sugar Maple, and American Beech, averaging 350 years old and situated on flat ground. The tall canopy was thick and complete, and there was surprisingly little undergrowth in the deep shade cast by the great trees. It was quiet and peaceful, with strains of birdsong adding to the cathedral-like experience. The families walking the trail spoke in low tones, respectful of the ancient forest. Of course, the familiar Red-eyed Vireos sang from the deciduous edge of the old growth. Maples dominated the deciduous trees, with sapling American Beeches filling in the understory and a few mature beeches scattered through the tract. I heard Black-throated Green Warbler, Rose-breasted Grosbeak, Eastern Wood Pewee, and Pine Warbler. Off in the distance was the drum of the Pileated Woodpecker. Wouldn't it be great, I wondered, if this ancient forest encompassed 490 or 4,900 acres rather than just 49?

The museum told not only the logging story but those of its victims: wild nature and the poorly paid loggers who brought down the great timber. It was both riveting and depressing to examine closely the crisp black-and-white prints of the old logging scenes, accompanied by text enumerating the wholesale pillaging of this seemingly limitless resource by rapacious timber barons. These men cared not a whit for the proper treatment of their employees or the proper management of the vast natural resource that made them immensely wealthy. That first harvest of more than 19.5 million acres of virgin timber, none replanted by the greedy plunderers, permanently impoverished the upland forests of Michigan. Generating more than a billion board-feet of timber a year during the peak of operation, around 1890, the lumber produced more wealth than the California Gold Rush.

White Pine was the tree that made men rich, and most timber operations targeted it in particular. Only later, when the White Pine was gone, did other species gain in importance as marketable timber. The ancient forests that the loggers found when first cruising the

Black-throated Green Warbler

wilds of northern Michigan were rich in conifers: not only the White Pine but also Red Pine, Eastern Hemlock, and White Spruce. Today these lands are dominated by maple, birch, and beech. The ecological and economic differences between the ancient mixed conifer forests and today's deciduous forests are considerable. The impacts this forest destruction has had on the wildlife and overall biodiversity of the state have been measureless.

Given the breathtaking avarice of the timber barons, it is a miracle that the remnant forty-nine acres at Hartwick Pines were never cut. The objective of the timber companies was to harvest every merchantable tree. Today, the most productive forest lands in Michigan have been cut over three times, and true old-growth forest lurks in only a few protected areas. The best are in the Upper Peninsula: the thirty-one-thousand-acre Porcupine Wilderness State Park and the eighteen-thousand-acre Sylvania Wilderness of Ottawa National Forest.

Both large old-growth tracts lie a few hours east of where I camped in northern Wisconsin, but, sadly, I had not known of their existence when I passed through the area.

BACK NORTH

After the bittersweet experience of Hartwick Pines, I head north on Inter-state 75 toward the Trans-Canada Highway and Chutes Provincial Park, in Massey, eastern Ontario, 150 miles east of Sault Sainte Marie. It is a stunning late-spring Sunday afternoon. I stop for lunch on the northern side of the Mackinac Bridge, at Saint Ignace—a lovely tourist destination with unrestricted views of the water. It has the look of a beach town, tailored to summer tourists, a bit like Newburyport, Massachusetts, but without the seafaring history. I dine al fresco as a light breeze comes off the straits. All around me families with children celebrate their weekend amid the vacation vibe of this small shore town on a day of wondrous weather. On any given day, it could just as well be 58°F here, with rain and twenty-mile-an-hour winds off the straits. We are all thankful for the meteorological bounty that has come our way.

Along much of the Trans-Canada Highway, I had scenic watery views over the North Channel, a large embayment of Lake Huron. The lake's north shore has a north-country look, with a mix of aspens, birches, spruce, fir, and Tamarack. An Osprey carrying a large fish passed over the car, followed by a Great Blue Heron. This shoreline is very much defined by its islands and peninsulas. For the first time on the trip, I saw a Banded (or White) Admiral, a butterfly I'd looked for since I reached Wisconsin. I'd seen not a one in northern Ontario, which remains a mystery.

Chutes Provincial Park is small—just 270 acres—but it contains the falls and rapids of the River aux Sables, and there are good hiking trails in its woods. After the long drive from Michigan, I was pleased to sleep in the next morning. An Ovenbird sang loudly beside my tent early, but I rolled over. It was good sleeping weather: the 8 a.m. temperature was 58°F. A four-mile hike on the Twin Bridges Trail began my day, and I found, in the middle of the trail, the severed back half of a Least Weasel (which proved to be the only weasel I encountered on my journey). I guessed the front half had been consumed by an owl, which had dropped the back half by some mishap.

It promised to be a warm and sunny day. Here *Trillium, Clintonia,* Bunchberry, and False Lily of the Valley had mostly finished flowering, though Orange and Yellow Hawkweed still bloomed. Southern Ontario had followed the deficient timber management practices I'd learned about in Michigan, and I could see its results here. This rocky, scrubby forest was very much a product of the late-twentieth-century timber overharvest, with young stands of Paper Birch (its abundance an indicator of fire as well as logging), White Pine, Red Pine, Jack Pine, Red Maple, aspen, and Red Oak, as well as small stands of fir, spruce, and Northern White-Cedar. The pretty blackwater river had a number of rocky chutes and rapids, although these are not the source of the park's name. Instead it was named after an old timber-era log chute that remained here as late as 1960—testament to the long run of logging along this watershed. Each spring, logs were sent down the wooden chutes to the mill in Massey. A Hermit Thrush sang from a thicket at 9 a.m., with the sun high, and Ovenbirds and Scarlet Tanagers sang from the woods. I saw evidence of a lot of recent timber cutting by a local beaver family.

Before midday, I was back on the road, headed about three hours farther east to Algonquin Provincial Park. I passed through the large communities of Sudbury and North Bay and turned south at Eau Claire onto a bucolic back road ending at Kiosk Lake, the park's northwestern entrance. A Broad-winged Hawk soared over the road as I passed through Nippising First Nations territory, marked by a large signboard. The Kiosk Road led south through a mix of landscapes, boreal spruce forest and north-country farms following the valley of the Amable du Fond River. History here, too, is all about logging—a recurrent theme in these parts.

I camped on Kioshkokwi Lake (Kiosk for short), a very minor corner of famed Algonquin Provincial Park, and I planned to spend only a night here. My time on the road was coming to an end, and I was headed rapidly east and south to my final destination, northern New York State. Exploring by bike, I found the usual cast of north-country characters: Common Raven, Blue Jay, Belted Kingfisher, Yellow-bellied Sapsucker, Eastern Kingbird, Ruffed Grouse, Red-eyed Vireo, and American Redstart.

The next morning I awoke to the sound of rain on my tent fly. A nasty weather system had come in overnight, and even at 9 a.m., I was still in the tent, waiting for the rain to break. When I emerged, I found the low cloud and steady light drizzle typical of the north country. Today was July 1—Canada Day. The weather wasn't promising for the annual celebration, which I planned to observe at the Black Bear campground in Petawawa, Ontario. Yet in spite of the rain, a male Magnolia Warbler sang from a branch just overheard.

I continued east through three hours of gloom and rain on the two-lane blacktop that was the Trans-Canada Highway. Petawawa sits on the south bank of the mighty Ottawa River, and Black Bear campground is down in a patch of woods on the shore of the river, which at first I took for a large lake—it is *that* broad. The campground was filled with Canadian families here to celebrate their nation's birthday.

This was a far cry from wilderness, but it was the best I could find. Naturizing yielded an Ox-eyed Daisy as well as a Beaked Hazel, a shrub I had seen in several spots in Minnesota. False Solomon's Seal was producing green berries, and Spreading Dogbane was flowering. Birdwise, I found a Pine Warbler and a Yellow-bellied Sapsucker, but not much else. I went to bed early, listening to earth-shattering fireworks shooting out over the river after 10 p.m. on this, my last night in Canada.

LEAVING THE NESTING GROUNDS FOR THE COAST

Many wood warblers nesting in Ontario finish breeding by midsummer. After raising their fledged young, the adults and the first-year birds separate, and both take a late-summer vacation before heading toward the Tropics. During this loafing period, the birds molt their feathers and start to fatten up for their long autumn travels. The adults generally depart before the young born that year. Of course, the adults already have experienced southbound migration and have first-hand knowledge of their route.

After departure, many of the birds that nest in the boreal forests of central Canada head eastward to the Atlantic coast rather than due south. They follow much the same route that I was traveling as I headed toward the Adirondacks in upstate New York. Their annual travel describes a loop—in spring, they head north up the Mississippi to Ontario; after breeding, they move east to the coast and then south to the Tropics. When they reach the East Coast part of the trip, many species spend time putting on fat before heading southward in earnest. This is a little-known period in the lives of these birds. They remain silent. They wear their drab autumn plumage. They stay hidden among the changing leaves.

NINE

Adirondack Spring

Early July 2015

Now, except for the stragglers, all the birds were back that were coming

back. Barn Swallows, home from Brazil, skimmed over the northern

pastures. Yellow Warblers, home from the Yucatán, darted along the

roadside. Bobolinks, home from Argentina, sang on the fences. The

great spring migration was over.

—EDWIN WAY TEALE, *North with the Spring*

At 6 a.m., an American Robin sings and a Pileated Woodpecker drums in the sharp postdawn cold. I depart overcrowded Black Bear Campground, traveling southeastward toward Ottawa. I am surprised to see a Brown Thrasher—a bird of the South—fly across the road, and then a Green Heron. Barn Swallows twist and turn over an large old field, where male Bobolinks hover improbably above the tall grass. I cross a small river, labeled "Mississippi," west of Ottawa. Both river and road cut through handsome, thick beds of pale-gray limestone before passing right through Ottawa's crowded western suburbs. I struggle to find Route 416 toward the New York border at Ogdensburg. A couple of days earlier, I'd been on the wild northern shore of Lake Huron, but now I am in flat farming country and the suburbs of Canada's national capital. I am in a hurry to get back to wilderness—the interior of the Adirondack Mountains.

On my reentry to the United States, I took a high bridge across the broad and deep blue Saint Lawrence River. I told the welcoming U.S. border control officer that I was an ornithologist on a field trip, and he pointed up to a pair of Ospreys nesting on a platform. From Ogdensburg, I navigated a maze of back roads toward the six-million-acre Adirondack Park, the largest tract of wild lands in the eastern United States. I'd spent twelve summers at camp in these mountains in the 1960s and '70s, and this was my long-overdue homecoming.

There were more welcoming sights and sounds along the two-hour back-roads drive through upstate New York. I saw the Eastern Bluebirds common in the Saint Lawrence Valley. Northwest of the park, the land is rural and open and agricultural, with only the slightest rolling hills. Eastward, billowing white clouds obscured the foothills of the Adirondack Mountains. I passed a bearded Amish man in a hat, blue shirt, and suspenders, guiding a horse-powered wagon. Next I slowed to pass a big Amish cart loaded with firewood and pulled by two draft horses, with two adorable children in the back

---

OPPOSITE: Bobolink

and a man in standing in front to drive the horses along. Crossing into the park at 10:30 a.m., I sped along the winding two-lane roads of this lake-filled, mountain-bedecked forest wilderness and passed through Lake Placid, twice host of the Winter Olympics. I turned south from the tiny community of North Elba onto Heart Lake Road, which offers one of the most picturesque mountain vistas in all the East, looking up to Indian Pass, Algonquin Peak, Mount Colden, and Mount Marcy, which together offer some of the finest mountain hiking in the Adirondacks. This was the ideal place to end my search for spring and to bag my last two breeding wood warblers.

JULY FROST

I set up my tent in the hemlock-shaded campground of the Adirondak Loj, right on tiny Heart Lake in the bosom of the High Peaks. As I worked, a Ruby-crowned Kinglet in a young Eastern Hemlock greeted me at eye level. A Blue-headed Vireo sang his syrupy song in a maple. And a female Purple Finch perched at the very pinnacle of a mature spruce.

In the afternoon, I hiked a steep and rocky trail to the summit of nearby Mount Jo, which looks over Heart Lake to the highest summits in the mountain wilderness. The half-hour climb granted a superb view of the best of the Adirondacks. Sitting atop a big, flat block of anorthosite granite, I was content to be in this spot at this time, close to the end of my cross-country adventure. Abby Katsos, an Adirondack Mountain Club intern, greeted me; she spent the days on Mount Jo to educate arriving hikers about wise treatment of the mountain summit vegetation and to answer questions about the park and its natural history. Katsos, from Pittsburgh, was interested in animal tracking. During my hour on Mount Jo, I was visited by Cedar Waxwings, a Myrtle Warbler, and a Dark-eyed Junco. Up here, the Balsam Firs were fruiting like mad, each with an abundance of erect deep-blue cones. When these cones mature and open, groups of winter finches will descend upon them, competing with Red Squirrels for the dining bonanza.

Looking south from North Elba to the Adirondack High Peaks

On my hike down to camp, I stopped at the Loj's nature center and met Heart Lake's three summer naturalists (what a wonderful place for them to spend the summer). Together we talked about breeding birds and speculated on where I might find a Brown Creeper—a common species that had eluded me over the three months of my journey. The naturalists knew their stuff and gave me hints on where to search.

The Adirondak Loj, owned by the Adirondack Mountain Club, serves as the club's center of summer operations. The Loj grounds were once the site of a woodland retreat built by Henry Van Hoevenburg, who named Heart Lake and Mount Jo (the latter for his fiancée, Josephine Scofield). The original Loj burned in a forest fire in 1903, and the current, smaller structure was built in 1927. Today it provides food and lodging for hikers and campers year-round and is a beloved destination in the High Peaks region of the great park—there's nothing quite like completing a big climb and returning to the Loj for a hearty meal and a good night's recuperation, with Barred Owls hooting into the night.

Why is it *Adirondak Loj* and not *Adirondack Lodge*? Melvil Dewey, president of nearby Lake Placid Club, purchased the Loj from Van Hoevenburg in 1900. Dewey, an American librarian and educator and inventor of the Dewey Decimal system of library classification, was also a supporter of a shift to simplified spelling of American English words. He put his belief into practice, condensing his first name from the traditional Melville as well as eliding the *c* in *Adirondack* and transforming the *dge* of *Lodge* into a simple *j*. Dewey, Theodore Roosevelt, Mark Twain, and Andrew Carnegie created the Simplified Spelling Board, which sought to make English simpler, phonetic, and thus more palatable as a global language. Roosevelt and Dewey used this spelling in their correspondence, but it was a difficult sell to Congress and the public, and it resulted in only minor changes to our common language (e.g., *plow* instead of *plough*; *honor* instead of *honour*).

The Adirondack Mountain Club, founded in 1922, today has twenty-eight thousand members and twenty-seven chapters across New York State. The club is among the powerhouse advocacy groups fighting to preserve the wilderness values of the Adirondacks and promoting the educational value of the wilderness experience for young and old alike. The club maintains and restores hiking trails, protects and restores sensitive alpine plant communities on high summits, teaches outdoor skills, and offers guided hikes and adventures. It works closely with the New York Department of Environmental Conservation to ensure far-sighted management of Adirondack Park.

July 3, my first morning in the Adirondacks, was cold enough to require gloves and wool cap at my campsite. The dawn broke with exuberant song by American Robins and Red-eyed Vireos, and I biked out to South Meadows in search of birds. The meadows sparkled under a coating of frost in the early sunlight: the very last breath of boreal spring. A Swainson's Thrush sang from a thicket of Balsam Fir. A sapsucker drummed in the distance. A Black-capped Chickadee, Red-breasted Nuthatch, Blue-headed Vireo, Ovenbird, Common Yellowthroat, and Magnolia Warbler welcomed me to the meadow.

Then I walked the trail circumnavigating Heart Lake in search of a Brown Creeper. The preceding afternoon, one Loj naturalist had pointed me to this trail as a good spot for this retiring species. I found thickets of Hobblebush, a big stand of old Sugar Maples, and a

Snowshoe Hare, but no creeper. Postwalk, I stopped for a hot breakfast at the Loj, served family-style at big rustic tables, and I had great fun chatting with the mix of guests who had come from all over.

Later in the morning, I drove north to the trailhead for Pitchoff Mountain, which looks across the narrow chasm of the Cascade Lakes to the summit of Cascade Mountain. I would climb Pitchoff rather than Cascade Peak because, on this Saturday in early July, the trail up Cascade likely would be thronged with hikers wanting to bag one of the "Adirondack 46" (the forty-six peaks in the park that are over four thousand feet tall). Pitchoff, a lesser mountain, would be quieter. The climb to its summit overlook was very steep, but once there I commandeered a section of rocky ledge, with the bright sun rising above the summits. The mountaintop conifers harbored singing Nashville and Myrtle Warblers, Red-eyed Vireos, White-throated Sparrows, and, of course, Swainson's Thrush. A surprise awaited me up here, too: a pair of Peregrine Falcons racing about through the narrow valley between the mountains. On a breathtaking fly-by, one raptor's wings made a tearing noise as they sliced through the chilled late-spring air.

On July 4, the morning's first birds included a Wild Turkey, a Ruffed Grouse, a sapsucker, a Ruby-throated Hummingbird, and—new for the trip!—a Brown Creeper. I located the shy little bird when I heard its tinkling territorial song in a dark thicket of hemlocks.

After biking down to the Loj for the hot breakfast, I headed off to climb Mount Van Hoevenburg. The trail from South Meadows travels through an old Red Pine plantation and a large beaver swamp, and then climbs gradually up the back of the mountain, which is home to the Olympic bobsled run on its north side. From the summit ledge on the back of the mountain, which is a northern outlier, one can look south to the whole panorama of the Adirondack High Peaks, which were cloud-free this morning. On the summit, I was serenaded by a confiding Winter Wren, a Blue Jay, a White-throated Sparrow, and a Myrtle Warbler. On the hike down, I came upon several noisy sap-suckers, a Black-throated Blue Warbler (a quest bird!) in the deciduous growth, and a Mourning Warbler in the beaver swamp. The day ended with a thunderstorm, which I waited out in my car, cooking dinner late, after the rain had cleared off. I had finished reading in my tent when loud reports began to echo down the valleys from Lake Placid town: Independence Day fireworks. They seemed to go on forever. I was far enough away that I could drift off, with thoughts of my big climb planned for the following morning.

July 5 broke cool and fine. From a clearing I looked up to the MacIntyre Range and the moon above it, without a cloud anywhere. I would be headed up that way this morning. By 5 a.m., I was biking south to the trailhead for Marcy Dam. It would take me three hours to summit Algonquin Peak—at 5,115 feet, the second-highest mountain in the Adirondacks. Here I would complete my three-month journey, add my last quest bird (plus the last nonquest bird of the trip), and finish up with a live radio interview with Ray Brown on his show, *Talkin' Birds*.

The wooded low country was busy with birds this morning: Swainson's Thrushes, Ovenbirds, and Blue-headed Vireos. As I climbed the initial ridge, I found Bunchberry in fresh flower—I was retreating earlier into spring as I gained elevation. I was the first hiker on the mountain this morning. The trail was mainly sloping shield rock, with small streams coursing down the rough granite face in many places, and my hiking boots gripped it well as I moved higher and higher. I passed through a section that had been burned long ago; a nearly pure stand of aging Paper Birch was dying out and being replaced by Balsam Fir. By 7:30 a.m., I was in pure Balsam Fir, where I heard my first Blackpoll Warbler, a species I had seen in migration but whose boreal-montane breeding habitat I was now entering for the first time on the trip. It was a treat to get a close look at the black-capped, white-cheeked, yellow-legged male moving gingerly among the fir boughs, its high-pitched song—*ts ts ts ts ts ts ts ts ts*—barely discernible except at very close range.

Black-throated Blue Warbler

Proceeding up the mountain a few minutes later, I heard the thin, rapid musical phrases of Bicknell's Thrush from a nearby thicket of Balsam. It was species number 259 of the hundred-day field trip, and the last bird of my journey. This mountaintop will-o'-the-wisp, rarely seen in migration, is found in summer only in the spruce-fir zone of Northeast mountains and in winter in remnant tropical upland forests of Hispaniola and Cuba. Hearing this Neotropical migrant's reeling song is one of the treats of the Adirondack High Peaks. The singer had a black-spotted breast, olive-brown upper parts, and dark eyes, looking like a small version of the more common and widespread Swainson's Thrush. Both of these thrush species distinguish themselves by their lovely territorial songs. The Bicknell's upland forest wintering habitat on Hispaniola is under threat. Moreover, the many small patches of mountaintop habitat in the Northeast could lose their ability to host Bicknell's Thrush under the growing influence of climate change. The Vermont Center for Ecostudies, through its ongoing field studies, is working to develop programs to ensure the health of both wintering and summering populations of the thrush.

As I climbed higher, the firs became more stunted. Before long I was in the open, and the vast Adirondacks spread out in all directions—forests in various shades of green, dark lakes, and mountains of varying shades of green, blue, and purple, depending on distance. I reached the high, rocky summit at 8:20 a.m., sharing the heavenly perch with only birds and alpine wildflowers. The wind blew strongly from the northwest, and a thin mist passed over the rocky dome, chilling me to the bone. I pulled out my fleece and found a sheltered rocky niche in the lee of the summit to soak up the sun while being protected from the cutting wind.

A dwarf spring garden grew here: recumbent firs and spruce, no more than a foot or two high, clung to the rocky substrate. In the openings, dwarfed flowering plants hugged the gravelly soil—False Hellebore (liberally glazed with frost), *Diapensia*, Mountain Sandwort, Bunchberry, and *Clintonia*, alongside dwarfed versions of Bog Laurel, Labrador Tea, Blueberry, Mountain-ash, and other woody plants. Everything lay low, hiding from the desiccating effect of the relentless summit wind. Here the brief alpine spring was still only arriving, soon to be followed by a short summer before the advent of the long autumn and winter. I had finished my journey in a high place where spring arrives in early July, a hundred days after my start in the Deep South.

As I rested and let the sun hit my face, I heard the tinkling trill of a snowbird—a Dark-eyed Junco, singing from a perch in a dwarf Balsam Fir. A minute later a raven soared overhead and gave a single low croak, his wings grabbing the stiff breeze. A Myrtle Warbler, one of the migrant songbirds I'd followed from Texas up the Mississippi and into Ontario and New York, gave its sad, musical, rambling trill from the thick carpet of firs downslope, reminding me of my first climb of this very peak in July 1965, when I was thirteen years old: a half century ago.

LAST OF THE BREEDING WARBLERS

After three months of effort, I had seen all the East's breeding wood warblers on their breeding habitat, from Louisiana to Minnesota and from Ontario to the Adirondacks. The last two species (Black-throated Blue and Blackpoll Warblers) had drawn me back here to the mountains, and in many ways both are iconic wood warblers—well known, beloved, and widespread. They breed in the North Woods, wintering in the Caribbean, Central America, and northern South America.

The Black-throated Blue Warbler (quest bird number 36) is a favorite of many birders. It is a common breeder in mature mixed forests of the Adirondacks. There are few more handsome birds; the male is simply but elegantly plumed in slate blue, black, and white. It forages in the understory, so it is visible at eye level, and the male's buzzy song, *zeeoah zeeoh zeee?*, alerts the birder to its presence. The bird is among the easiest wood warblers to track down in the forest. It is an eastern species, breeding in the Appalachians and upland New England, but is essentially absent as a breeder from the Mississippi drainage. That's why I was seeing it on its breeding habitat for the first time here; in fact, I had first seen the bird on its breeding grounds in the central Adirondacks back in 1965.

The Blackpoll Warbler, a widespread migrant and the last of my quest warblers, prefers mountain summits of balsam fir as a nesting habitat. That's why I had climbed Algonquin Peak, whose upper slopes are a sure place to find the bird on its late-spring breeding territory. The Blackpoll traditionally has swept through the woods of the East Coast on its way north, its passage announcing the end of spring migration. It holds all sorts of memories for those who love warblers, yet sadly, in the past two decades, its numbers have declined substantially. I recall encountering thirty or more on a spring morning in Washington, D.C., but now I rarely hear or see more than one or two in a day. It is alarming when a commonplace species becomes scarce. What has gone wrong?

THE ADIRONDACKS—THEN AND NOW

What I saw in my five days in the Adirondacks reminded me of what I recalled from the 1960s and 1970s. I was unable to detect major changes to the wild lands or the built environment; I visited the towns of Tupper Lake, Saranac Lake, and Lake Placid, and all were still picturesque, compact, and pleasingly old-fashioned. Perhaps this is not surprising, as the population of the Adirondacks has increased very little since 1900. At that time, it was 100,000, and in the year 2000, it was still only 130,000. The Adirondack Park Agency (APA), created in 1971, has successfully prevented the overdevelopment of this great forest reserve. The goal of the APA is to balance environmental concerns with economic development, and thus there has been concentrated development in centers of tourism (such as Old Forge and Lake Placid) while increasing amounts of wild land has been bought by the state and placed under strict protection. Again a comparison: in 1900, there were 1.2 million acres of state-owned land. This number had nearly tripled by the year 2000, while the number of sawmills in the park dropped by 50 percent in the same period.

Still, some negative trends are apparent, such as the advent of acid rain and climate change. The former is caused by sulfuric and nitric acids produced by the combustion of fossil fuels (sulfur from coal, nitrogen from coal and gasoline). These pollutants from power plants, industries, and automobiles travel hundreds of miles from their sources and fall with rain into the lakes and high-elevation conifer forests of the Adirondacks. With its high levels of precipitation, impermeable bedrock, and relatively high elevations, the Adirondack region has been particularly vulnerable to acid rain, which releases aluminum from soils. The aluminum and the acid itself kill fish and other aquatic species. Luckily, the Clean Air Act has reduced the impact of acid rain over the last decade or so. As a result, the population of an iconic Adirondack bird—the Common Loon—has rebounded to healthy levels. I did not see evidence of acid rain in my hike up Algonquin, but it historically has damaged conifers on certain summit forests and impacted lake fisheries. There has been progress, but this is a long-term phenomenon, requiring a long-term solution. Our government should not give up on the fight for cleaner air and cleaner water.

Climate change, too, is impacting the Adirondacks. I did not notice its effects in my visit, but it has had impacts on the birdlife. For instance, two of the specialty boreal bird species I encountered in northern Ontario, the Spruce Grouse and the American Three-toed Woodpecker, have, since the 1960s, declined to the point of disappearance in the Adirondacks. Most experts assume the demise of the local breeding populations of the two species is the result of climate change. Just what physical or biotic stressors are harming these two special birds is unknown, but the declines appear real.

Changes in the area's winter birdlife have been well documented in a paper by Larry Master that analyzes the results of the Christmas Bird Count over a sixty-year period in the central Adirondacks. More than a dozen bird species absent from early counts have become commonplace in more recent counts, including the Canada Goose, Hooded Merganser, Common Merganser, and Great Blue Heron—all of which require open water for foraging. Now Adirondack lakes freeze later in the autumn and melt sooner in the spring. Today the Adirondacks see much reduced snowfall and many fewer severely cold winter days than they did fifty years ago.

So, acid rain and fewer cold and snowy winters are serious changes in the Adirondacks. How these ongoing chemical and meteorological changes shall further alter the region is unknown, but additional biotic impacts may show themselves in decades to come. The world is changing, and our favorite places will change with it.

TROPICS BOUND

The last stage of this story of migrant songbirds takes place here in the Adirondacks. The warblers and vireos that nest in June and early July find themselves in a holiday feasting period in late July and August and into September. During this time, adults and hatching-year birds seem to follow distinct courses, though only pieces of this story have been delineated. Adults search out productive staging areas where they can forage on rich food resources, and probably young birds do the same thing, though apparently not in the same locations. The two age groups also appear to take different pathways to their winter homes in the Tropics. Both groups do seem to drift eastward in the last weeks of summer, heading toward New England, but long-term bird-banding operations in New England show that adult songbirds tend to concentrate in rich interior forest areas, while young birds tend to end up on the coast. Recall that older birds have experience with this southbound trip, whereas younger ones do not and so must follow instinctual impulses.

We do know that two of our focal species, the Connecticut and the Blackpoll Warblers, carry out a remarkable southbound journey as adults. Both species move east from their breeding habitat to the coast of New England (Blackpoll) or the Mid-Atlantic (Connecticut) and then depart southeastward out over the ocean, making a nonstop flight to the northern coast of South America. Some small number of individuals of both species touch down in Bermuda, which is on the flight path southward. The hardiest individuals that make the whole overwater trip fly nearly two thousand miles—a flight that might take as much as eighty hours. This heroic journey requires incredible reserves of energy, which is why the birds spend weeks bulking up in the interior of New England in early autumn before they make the flight. The Blackpoll has been shown to double its weight prior to departure from New England so that it will have the necessary energy stores for its trip. Earlier in this book, I discussed migrant songbirds' northbound flights across the Caribbean and the Gulf of Mexico. Yet even those flights pale in comparison to this autumn trek to South America. It is the most amazing feat of any Neotropical songbird migrant that we know of to date—the mind-boggling end product of the process of organic evolution operating on tiny songbirds over tens of thousands of generations.

EPILOGUE

Here in this wild and beautiful spot amid the mountains, the dark woods, the rising mist, the new moon hanging above the silhouettes of the peaks, we waited, in spite of the night chill, until the last sunlight of the spring had ebbed from the sky.

—EDWIN WAY TEALE, *North with the Spring*

From my Adirondack campsite, I drive home, heading into the glossy blue-gray skies of the hot Mid-Atlantic, where cicadas drone in the afternoons and pop-up evening thunderstorms douse suburban yards. I spend my last night on the road camping at Shawnee-on-Delaware, at the Water Gap in eastern Pennsylvania, where I listen to the songs of summer birds—Indigo Buntings, Yellow-throated Vireos, and Great Crested Flycatchers. Spring is done, finally, for me.

I think back to my brief stay on the summit of Algonquin Peak, where I'd called in to chat live with Ray Brown on his Sunday-morning radio show, Talkin' Birds. *It had given me a chance to reminisce about my three months on the road, driving 12,891 miles, biking several hundred miles, kayaking several dozen more, and walking some unknown but substantial distance. I'd encountered 259 species of birds, including all 37 quest warblers, along the way. I told Ray's audience that my trip was all about the nature of "place." The long road trip had brought me to an array of marvelous southern destinations, led me to a lonely Jack Pine forest in Canada, and ended in a place filled with treasured childhood memories. Many of the southern locales were special for their novelty. The sojourn in northern Ontario took me to a distant land that fell short of my high expectations but nonetheless surprised and educated me. The last place I visited harkened back to my past. It was fine to end my journey in the Adirondacks—as wild and bountiful as it had been when I spent my first summer there as a twelve-year-old.*

P lenty of writers out there have cataloged the environmental losses we have suffered since the 1960s. There's no denying that we have forever lost some of our precious natural heritage over the ensuing decades. At the same time, however, there have been conservation successes, both small and large, and we have discovered more about nature's capacity to restore itself. Today, after all, there is more forest land under conservation in the Adirondacks than there was in 1964, when I first visited. That's good news. We must remain hopeful and continue to work to conserve and restore what we love.

PREVIOUS: Swainson's Thrush

North on the Wing

My wondrous experiences in the Tensas Basin, Delta National Forest, Wyalusing State Park, Crex Meadows, and Pukaskwa National Park showed me that North America remains filled with unique green spaces worth visiting and protecting. Seeing a singing Connecticut Warbler in northern Ontario was fantastic but no less special than seeing a singing Bicknell's Thrush on Algonquin Peak. There are many tantalizing bird encounters to be had on this broad continent, and we can all be out exploring our favorite old haunts and discovering new destinations near and far. It makes our souls sing, and it keeps us young at heart.

On the nature conservation front, there is work to be done. For the beloved Neotropical migratory songbirds, the challenge is clear: we must pull together to conserve, restore, and expand critical blocks of forest habitat that the birds require for their breeding territories, their winter homes, and their various stopover sites in between. These blocks of habitat should be doubled or tripled in size. If we succeed in this big task, we can reverse the decline of our songbirds and bring them back to the abundance celebrated by Edwin Way Teale in his 1951 book. We need to throw our support behind institutions working on behalf of nature in North America—ones such as the Houston Audubon Society, the American Bird Conservancy, the Texas Parks and Wildlife Department, the Smithsonian Migratory Bird Center, and the U.S. Fish and Wildlife Service, to name only a few.

In many places, we have moved in the right direction. Think again of the Adirondacks: when young Theodore Roosevelt first did his bird studies there in the 1870s, not a single acre had been protected. In 1885, as governor of New York, Roosevelt created the Adirondack Preserve—a first step. Coincidentally, Roosevelt, by then vice president, was climbing Mount Marcy in the Adirondack High Peaks in 1901 while President William McKinley lay in the hospital, felled by an assassin's bullet; McKinley soon died, making Roosevelt president. Roosevelt's love of the Adirondacks and other wild places in America inspired him to create the first national parks, based on the Adirondack model.

White-throated Sparrow

At the time of Roosevelt's presidency, many private forest tracts in the Adirondacks were being heavily logged, and large-scale timber felling was happening in Michigan, Wisconsin, and Ontario. Today many of the areas logged in the early 1900s are classified as protected forest, never to be logged again. Yes, the ancient forests—the virgin woods—are mostly gone from the East, but the expansion of mature forest cover throughout the East has risen relentlessly since the 1930s. Today its extent is considerably greater than it was in the year 1800. Wildlife species that were completely extirpated from the East, such as the Wild Turkey, Fisher, and North American Beaver, are now back in force, settling in suburbs and visiting backyards. Let's never forget Nature's regenerative capacities. Were Edwin Way Teale alive today, he would be amazed at the regreening of the East, especially in and around urban areas. That said, he probably would express worry over the future of the Neotropical songbird migration phenomenon. He would be surprised by the impact of human demography on the forests of Central America, northern South America, and the insular Caribbean. He would, I believe, assert the need for every citizen to take notice and to work to conserve or restore those special green spaces that still exist. Reaching out to our compatriots south of the U.S. border is something we need to do more, sharing best practices and helping our counterparts in the Tropics to see the biotic linkages that migration produces. They must conserve their backyard as we continue to conserve ours. Organizations such as the American Bird Conservancy are working with partners to bring about full-life-cycle conservation of migratory birds. Conservation biologists such as Pete Marra and his research team at the Smithsonian are continuing to ask the questions that help us to better understand the mechanics of migration and the particular points of vulnerability that need shoring up. Such understanding will inform the actions we'll take to reduce the threats posed by lighted tall buildings, transmission towers, wind turbines, and free-roaming cats—as well as the more familiar threat of habitat loss.

I remain hopeful. This hope may come with age and experience; comparing my encounters with wild nature in the Mid-Atlantic region in the 1960s to those I've had today is enlightening. In the spring of 2016, a family of five Common Ravens (our largest songbird) spent several weeks in my suburban neighborhood, adjacent to the Potomac River and Washington, D.C. By contrast, in the 1960s, a raven was a very rare sight anywhere east of the high, rocky ridges of the Appalachians. In the same vein, today Bald Eagles nest along the Potomac River and pass overhead in and around the nation's capital on a regular basis. I counted more than twenty-five of them on a recent Christmas Bird Count on the Virginia side of the mid-Potomac. By contrast, a Bald Eagle would have been a rare sight indeed along the Potomac in the 1960s. The same story can be told of the Osprey, the Peregrine Falcon, and a number of other once imperiled species.

I see reason for hope, but I also make a plea for vigilance on behalf of wild nature. More citizens' action, undertaken by more of us working in concert, can lead us to a new and better world. Let's all raise the flag for wilderness, wood warblers, and all things wild.

ACKNOWLEDGMENTS

I offer thanks to those who made my journey possible. My mother got me started in the nature business. My brother, Bill, became my first fellow naturalist—we saw some great birds and collected some awesome butterflies in our day. We have never forgotten seeing our first flock of Evening Grosbeaks in 1962. Martha Schaeffer and various other ladies of the Maryland Ornithological Society, with its little museum on the third floor of the Mansion at Cylburn Park, offered weekend nature study and kept interest growing in many youngsters, myself included.

Many advised and aided me in planning and executing the 2015 road trip. A number of beautiful green places that I visited had to be excluded from this narrative to keep the book compact. I apologize to those who kindly aided me in those wonderful places that, sadly, are no longer featured in the story. I wish I could have included every place I visited.

Here are those who assisted on the field trip and who aided preparation of the manuscript: Vanessa Adams, John Anderton, Isaac Betancourt, James Brown, Mia Brown, Ray Brown, Emily Cohen, Nathan Cooper, Marie Correll, Evan Dalton, Jeff Denman, Aditi Desai, Don DesJardin, Peter Dieser, Scott Ellery, John Faaborg, Adele Faubert, George Fenwick, Rita Fenwick, Tracy Fidler, Gabriel Foley, Susan Fulginiti, Steven Goertz, Mark Guetersloh, Tim Guida, Christian Hagenlocher, Larry Heggemann, Bridget Hinchee, Marshall Howe and Janet McMillen, Rich Kazmaier, Erin Lebbin, Bobby Maddrey,

Pete Marra, Todd Merendino, Frank Moore, Merrie Morrison, Musselwhite Mine, Gary Neace, Clare Nielsen, Mike Parr, Kelly Purkey, Kacy Ray, Nathan Renick, Mark Robbins, Paul Schmidt, Cliff Shackelford, Don Sisson, Lela Stanley, Keegan Tranquillo, Kinney Wallace, and Sister Mary Winifred. The American Bird Conservancy sponsored this adventure. And nine close friends—you know who you are—generously underwrote the costs of the field trip reported on here. They, along with a grant from the Georgia-Pacific Corporation to the American Bird Conservancy, made my pilgrimage possible. Roger Pasquier, Jared Keyes, Laura Harger, Grace Beehler, Carolyn Gleason, and Dominic Nucifora provided useful editorial direction to my revisions of the original draft. Bill Nelson prepared the endpaper maps. Carol Beehler designed the book and advised me on various issues. To these and others, I offer my gratitude.

REFERENCES

Works are listed in those chapters in which the works are first quoted or to which they are most relevant. Works are not relisted in subsequent chapter sections, even when quoted in or pertinent to those chapters.

INTRODUCTION

Bird Conservation magazine. Various issues. The Plains, VA: American Bird Conservancy.

Carson, Rachel. 1962. *Silent Spring*. Boston: Houghton Mifflin.

Halle, Louis. 1947. *Spring in Washington*. New York: Harper and Brothers.

McPhee, John. 1998. *Annals of the Former World*. New York: Farrar, Straus and Giroux.

Peterson, Roger Tory, and James Fisher. 1955. *Wild America*. Boston: Houghton Mifflin.

Sibley, David Allen. 2009. *The Sibley Guide to Trees*. New York: Alfred A. Knopf.

Stutchberry, Bridget. 2007. *Silence of the Songbirds*. New York: Walker & Co.

Teale, Edwin Way. 1951. *North with the Spring*. New York: Dodd, Mead and Co.
———. 1953. *Circle of the Seasons*. New York: Dodd, Mead and Co.

Wilcove, David S. 2004. *The Condor's Shadow: The Loss and Recovery of Wildlife in America*. New York: W. H. Freeman.

ONE | BIRDS OF SPRING

Chapman, Frank M. 1907. *The Warblers of North America*. New York: D. Appleton & Co.

Chu, Miyoko. 2007. *Songbird Journeys: Four Seasons in the Lives of Birds*. New York: Walker & Co.

Dunn, Jon, and Kimball Garrett. 1997. *Peterson Field Guides: Warblers*. Boston: Houghton Mifflin.

Lovette, Irby J., and John W. Fitzpatrick, eds. 2016. *The Cornell Lab of Ornithology Handbook of Bird Biology*, 3rd ed. Chichester, West Sussex, UK: John Wiley & Sons.

Mayor, Stephen J., et al. 2017. Increasing phenological asynchrony between spring green-up and arrival of migratory birds. *Nature/Scientific Reports* 7, no. 1902, doi:10.1038/s41598-017-02045-z.

Morse, Douglass H. 1989. *American Warblers.* Cambridge, MA: Harvard University Press.

North American Bird Conservation Initiative. 2016. *The State of North America's Birds 2016.* Ottawa, ON: Environment and Climate Change Canada.

Thoreau, Henry David. 1951. *Walden.* New York: Bramhall House.

Weidensaul, Scott. 1999. *Living on the Wind: Across the Hemisphere with Migratory Birds.* New York: North Point Press.

———. 2017. The new migration science. *Living Bird* 36, no. 2: 33–40.

Wilcove, David S. 2008. *No Way Home: The Decline of the World's Great Animal Migrations.* Washington, DC: Island Press.

Winkler, David, et al. 2017. Evolution of migration. *Living Bird* 36, no. 2: 24–29.

TWO | THE TEXAS GULF COAST: FIRST LANDFALL

Bub, Hans. 1991. *Bird Trapping and Bird Banding.* Ithaca, NY: Cornell University Press.

Ducks Unlimited. Undated. Home page. www.ducks.org/.

Horton, Kyle G., Benjamin M. Van Doren, Phillip M. Stepanian, Andrew Farnsworth, and Jeffrey F. Kelly. 2016. Seasonal difference in landbird migration strategies. *Auk* 133: 761–69.

National Wetlands Research Center. Undated. The coastal prairie region. www .nwrc.usgs.gov/prairie/tcpr.htm.

The Nature Conservancy. Undated. Home page. www.nature.org/.

———. Undated. Clive Runnells Family Mad Island Marsh Preserve. www .nature.org/ourinitiatives/regions/northamerica/unitedstates/texas/ placesweprotect/clive-runnells-family-mad-island-marsh-preserve.xml.

Niering, William A. 1987. *Wetlands.* New York: Borzoi Books.

Rockwell, S. M., C. I. Bocetti, and P. P. Marra. 2012. Carry-over effects of winter climate on spring arrival date and reproductive success in an endangered migratory bird, Kirtland's Warbler (*Setophaga kirtlandii*). *Auk* 129, no. 4: 744–52.

Smith, Gregory. 2013. *The U.S. Geological Survey Bird Banding Laboratory.* Reston, VA: U.S. Geological Survey.

Smithsonian Migratory Bird Center. Undated. Home page. https://nationalzoo .si.edu/scbi/migratory birds.

Wilson, S., S. L. LaDeau, A. P. Tottrup, and P. P. Marra. 2011. Range-wide effects of breeding- and nonbreeding-season climate on the abundance of a Neotropical migrant songbird. *Ecology* 92, no. 9: 1789–98.

Winger, Ben, et al. 2014. Temperate origins of long-distance seasonal migration in New World songbirds. *Proceedings of the National Academy of Sciences of the USA* 111, no. 33: 12115–20.

American Bird Conservancy. Undated. Home page. https://abcbirds.org/.

Anahuac National Wildlife Refuge. Undated. Home page. www.fws.gov/refuge/anahuac/.

Bowlin, Melissa S., et al. 2015. Unexplained altitude changes in a migrating thrush: long-flight altitude data from radio-telemetry. *Auk* 132: 808–16.

Dittmann, Donna L., Steven W. Cardiff, and Jay V. Huner. 2015. Of rice and rails. *Birding* 47, no. 2: 36–45.

Eubanks, Ted, Paul Kerlinger, and R. Howard Payne. 1993. High Island, Texas: case study in avitourism. *Birding* 25, no. 6: 415–20.

Farnsworth, Andrew, and Robert W. Russell. 2007. Monitoring flight calls of migrating birds from an oil platform in the northern Gulf of Mexico. *Journal of Field Ornithology* 78: 279–89.

Gallagher, Tim, Kristi Streiffert, and Sheila Buff. 2001. *Where the Birds Are.* New York: DK Publishing.

Gauthreaux, Sidney A. Jr. 1975. Radar ornithology: bird echoes on weather and airport surveillance radars. *NASA STI/Recon Technical Report* 75: 30397.

Gauthreaux, S. A. Jr., and C. G. Belser. 2003. Radar ornithology and biological conservation. *Auk* 120: 266–77.

Gauthreaux, Sidney A. Jr., John W. Livingston, and Carroll G. Belser. 2008. Detection and discrimination of fauna in the aerosphere using Doppler weather surveillance radar. *Integrative and Comparative Biology* 48: 12–23.

Graham, Frank. 1992. *The Audubon Ark.* Austin: University of Texas.

Houston Audubon Society. Undated. Home page. www.houstonaudubon.org.

———. Undated. High Island sanctuaries. www.houstonaudubon.org/default .aspx/MenuItemID/373/MenuGroup/High+Island.htm.

Lockwood, Mark W., and Brush Freeman. 2014. *The TOS Handbook of Texas Birds* 47. College Station: Texas A&M University Press.

Moore, Frank. 1990. Prothonotary Warblers cross the Gulf of Mexico together. *Journal of Field Ornithology* 61: 285–87.

———. 1999. Neotropical migrants and the Gulf of Mexico: the cheniers of Louisiana and stopover ecology. In K. P. Able, ed., *Gathering of Angels: Migrating Birds and Their Ecology*, 51–62. Ithaca, NY: Cornell University Press.

Moore, Frank, and Paul Kerlinger. 1987. Stopover and fat deposition by North American wood-warblers (Parulinae) following spring migration over the Gulf of Mexico. *Oecologia* 74: 47–54.

Shackelford, Clifford E., Edward R. Rozenburg, W. Chuck Hunter, and Mark W. Lockwood. 2005. *Migration and the Migratory Birds of Texas: Who They Are and Where They Are Going*, 4th ed. Texas Parks and Wildlife. https://tpwd .texas.gov/publications/pwdpubs/media/pwd_bk_w7000_0511.pdf.

Terborgh, John W. 1989. *Where Have All the Birds Gone?* Princeton, NJ: Princeton University Press.

Audubon, John James. 1960. *Audubon and His Journals*, vols. 1–2. New York: Cornell University Press.

Buchanan, Minor Ferris. 2002. *Holt Collier: His Life, His Roosevelt Hunts, and the Origin of the Teddy Bear*. Jackson: Centennial Press of Mississippi.

Durant, Mary, and Michael Harwood. 1987. *On the Road with John James Audubon*. New York: Dodd, Mead and Co.

Fitzpatrick, John W., et al. 2005. Ivory-Billed Woodpecker (*Campephilus principalis*) persists in continental North America. *Science* 308: 1460–62.

Gallagher, Tim. 2005. *The Grail Bird: Hot on the Trail of the Ivory-Billed Woodpecker*. Boston: Houghton Mifflin.

Heitman, Danny. 2008. *A Summer of Birds: John James Audubon at Oakley House*. Baton Rouge: Louisiana State University Press.

Jackson, Jerome A. 2004. *In Search of the Ivory-Billed Woodpecker*. Washington, DC: Smithsonian Books.

Meanley, Brooke. 1971. Natural history of the Swainson's Warbler. *North American Fauna* 69: iii–90.

———. 1972. *Swamps, River Bottoms & Canebrakes*. Barre, MA: Barre Publishers.

Roosevelt, Theodore. 1908. *In the Louisiana Cane Brakes*. New Orleans: Louisiana Wild Life and Fisheries Commission.

Snyder, Noel F. R., David E. Brown, and Kevin B. Clark. 2009. *The Travails of Two Woodpeckers*. Albuquerque: University of New Mexico Press.

Tanner, James. 1942. *The Ivory-Billed Woodpecker*. New York: National Audubon Society.

Theroux, Paul. 2015. *Deep South*. Boston and New York: Houghton Mifflin.

Ambrose, Steven E., and Douglas G. Brinkley. 2002. *The Mississippi and the Making of a Nation*. Washington, DC: National Geographic Society.

Bonney, Lorraine G. 2011. *The Big Thicket Guidebook*. Denton: University of North Texas Press.

Butcher, Russell D. 2008. *America's National Wildlife Refuges: A Complete Guide*. Lanham, MD: Taylor Trade Publishing.

Caro, Robert. 1990. *The Path to Power: The Years of Lyndon Johnson*. New York: Vintage.

Caroli, Betty Boyd. 2015. *Lady Bird and Lyndon: The Hidden Story of a Marriage That Made a President*. New York: Simon and Schuster.

Cobb, James C. 1994. *The Most Southern Place on Earth: The Mississippi Delta and the Roots of Regional Identity*. New York: Oxford University Press.

Critical Trends Assessment Program. 1998. *The Cache River Basin: An Inventory of the Region's Resources*. Urbana: Illinois Department of Natural Resources.

Dahmer, Fred. 1995. *Caddo Was . . . : A Short History of Caddo Lake.* Corrie Herring Hooks Series. Austin: University of Texas Press.

Dixon, James R. 2000. *Amphibians and Reptiles of Texas.* W. L. Moody Jr. Natural History Series. College Station: Texas A&M University Press.

Gabbe, A. P., S. K. Robinson, and J. D. Brawn. 2002. Tree-species preferences of foraging insectivorous birds: implications for floodplain forest restoration. *Conservation Biology* 16: 462–70.

Guelzo, Allen C. 2009. *Lincoln and Douglas: The Debates That Defined America.* New York: Simon & Schuster.

Hilty, Steve L. 2016. *Dirt, Sweat, and Diesel: A Family Farm in the Twenty-First Century.* Columbia: University of Missouri Press.

Monroe, Doug. 2001. *The Maverick Spirit: Georgia-Pacific at 75.* Old Saybrook, CT: Greenwich Publishing Group.

National Geographic Society. 2012. *Guide to State Parks of the United States.* Washington, DC: National Geographic Society.

Petrides, George A. 1972. *A Field Guide to the Trees and Shrubs.* Boston: Houghton Mifflin.

Robinson, Bob. 2008. *Bicycling Guide to the Mississippi River Trail.* Fort Smith, AR: Spirits Creek.

SIX | FROM THE CONFLUENCE TO THE HEADWATERS

Arthur, Anne. 2013. *Minnesota State Parks.* Cambridge, MN: Adventure Publications.

Burke, Alicia D., Frank R. Thompson III, and John Faaborg. 2017. Variation in early-successional habitat use among independent juvenile forest breeding birds. *Wilson Journal of Ornithology* 129: 235–46.

Cochran, William W., and Martin Wikelski. 2005. Individual migratory tactics of New World *Catharus* thrushes. In Russell Greenberg and Peter P. Marra, eds., *Birds of Two Worlds: The Ecology and Evolution of Migration,* 274–89. Baltimore: Johns Hopkins University Press.

Driftless Area account. https://en.wikipedia.org/wiki/Driftless_Area.

Ehle, John. 1997. *Trail of Tears: The Rise and Fall of the Cherokee Nation.* New York: Anchor Books.

Glassberg, Jeffrey. 1993. *Butterflies through Binoculars.* New York: Oxford University Press.

Holzschuh, Jennalee A., and Mark Deutschlander. 2016. Do migratory warblers carry excess fuel reserves during migration for insurance or for breeding purposes? *Auk* 133: 459–69.

Johnson, Charles W. 1985. *Bogs of the Northeast.* Hanover, NH: University Press of New England.

Leopold, Aldo. 1966. *A Sand County Almanac.* New York: Oxford University Press

Madson, John. 1985. *Up on the River: People and Wildlife of the Upper Mississippi.* Iowa City: University of Iowa Press.

Mathews, S. N., and P. G. Rodewald. 2010. Urban forest patches and stopover duration of migratory Swainson's Thrushes. *Condor* 112: 96–104.

North American Bird Conservation Initiative, U.S. Committee. 2014. *State of the Birds 2014 Report*. Washington, DC: U.S. Department of the Interior.

Palmer, Cynthia. 2013. Note to EPA: modernize tests you run on pesticides. *Bird Calls* 17, no. 2: 3.

Perich, Shawn. 2011. *Backroads of Minnesota*. Minneapolis: Voyageur Press.

Rea, Amy C. 2011. *Back Roads and Byways of Minnesota*. Woodstock, VT: Countryman Press.

Reid, Fiona A. 2006. *Mammals of North America*, 4th ed. Boston: Houghton Mifflin.

Sibley, David A. 2014. *The Sibley Guide to Birds*. New York: Alfred A. Knopf.

Silverberg, Robert. 1986. *Mound Builders*. Athens: Ohio University Press.

Smith, Robert J., and Margaret I. Hatch. 2017. Loss of Southern Arrowwoods (*Viburnum dentatum*) is associated with changes in species composition and mass gain by spring migrants using early successional habitats. *Wilson Journal of Ornithology* 129: 247–58.

Sutton, Ann, and Myron Sutton. 1987. *Eastern Forests*. New York: Borzoi Books.

Teale, Edwin Way. 1978. *A Walk through the Year*. New York: Dodd, Mead and Co.

Tekiela, Stan. 1999. *Wildflowers of Minnesota*. Cambridge, MN: Adventure Publications.

———. 2005. *Mammals of Minnesota*. Cambridge, MN: Adventure Publications.

SEVEN | THE MYSTERIOUS NORTHLANDS

Bertrand, J. P. 1997. *Timber Wolves: Greed and Corruption in Northwestern Ontario's Timber Industry, 1875–1960*. Thunder Bay, ON: Thunder Bay Historical Museum Society.

Borland, Hal. 1964. *Sundial of the Seasons*. Philadelphia: J. B. Lippincott Co.

Dashwood, Hevina S. 2014. *The Rise of Global Corporate Social Responsibility: Mining and the Spread of Global Norms*. New York: Cambridge University Press.

Heinrichs, Marj, and Dianne Hiebert. 2003. *Mishkeegogamang: The Land, the People and the Purpose: The Story of Mishkeegogamang Ojibway Nation*. Canada: Rosetta Projects.

Henry, J. David. 2002. *Canada's Boreal Forest*. Washington, DC: Smithsonian Institution Scholarly Press.

Imbeau, Louis, Jean-Pierre L. Savard, and Réjean Gagnon. 2000. Comparing bird assemblages in successional Black Spruce stands originating from fire and logging. *Canadian Journal of Zoology* 77: 1850–60.

Larsen, James A. 1982. *Ecology of the Northern Lowland Bogs and Conifer Forests*. New York: Academic Press.

Newton, Ian. 2008. *The Migration Ecology of Birds*. London: Academic Press.

Peterson, Roger T., and Margaret McKenny. 1968. *Wildflowers: Northeastern/ Northcentral North America*. Boston: Houghton Mifflin.

Rappole, John H. 2013. *The Avian Migrant: The Biology of Bird Migration*. New York: Columbia University Press.

Thoreau, Henry David. 1846. *The Maine Woods*. Boston: Ticknor & Fields.

Truth and Reconciliation Commission of Canada. 2015. *Honouring the Truth, Reconciling for the Future: Summary of the Final Report of the Truth and Reconciliation Commission of Canada*. www.trc.ca/websites/trcinstitution/ File/2015/Honouring_the_Truth_Reconciling_for_the_Future_July_23 _2015.pdf.

EIGHT | GREAT LAKES COUNTRY

Bothwell, Bob. 2007. *The Penguin History of Canada*. Toronto: Penguin Canada.

Cooper, N., M. Hallworth, and P. Marra. 2017. Light-level geolocation reveals wintering distribution, migratory routes, and primary stopover locations of an endangered long-distance migratory songbird. *Journal of Avian Biology* 48: 1–11.

Dennis, Jerry. 2004. *The Living Great Lakes: Searching for the Heart of the Inland Seas*. New York: St. Martin's.

Farb, Peter. 1968. *The Face of North America*. New York: Harper & Row.

Hargreaves, Irene M., and Harold M. Foehl. 1965. *The Story of Logging the White Pine in the Saginaw Valley*. Bay City, MI: Red Keg Press.

La Sorte, Frank A., et al. 2014. Spring phenology of ecological productivity contributes to the use of looped migration strategies by birds. *Proceedings of the Royal Society for Biology B* 281: 201409984.

Mayfield, Harold F. 1961. Vestiges of a proprietary interest in nests by the Brown-Headed Cowbird parasitizing the Kirtland's Warbler. *Auk* 78: 162–66.

———. 1977. Brood parasitism: reducing interactions between Kirtland's Warblers and Brown-Headed Cowbirds. In Stanley A. Temple, ed., *Endangered Birds: Management Techniques for Preserving Threatened Species*, ch. 11. Madison: University of Wisconsin Press.

McPhee, John. 1998. *Annals of the Former World*. New York: Farrar, Straus and Giroux.

Michigan Department of Natural Resources, United States Fish and Wildlife Service, and United States Forest Service. 2014. *Kirtland's Warbler Breeding Range Conservation Plan*. www.michigan.gov/documents/dnr/ Kirtlands_Warbler_CP_457727_7.pdf.

Morse, Douglass H. 1978. Populations of Bay-Breasted and Cape May Warblers during an outbreak of the spruce budworm. *Wilson Bulletin*: 404–13.

North on the Wing

Rapai, William. 2010. *The Kirtland's Warbler: The Story of a Bird's Fight against Extinction and the People Who Saved It*. Ann Arbor: University of Michigan Press.

———. 2016. Connecting community with conservation to secure a future for Kirtland's Warbler. *Birder's Guide to Conservation and Community* 28, no. 2: 14–17.

Waters, Thomas F. 1987. *The Superior North Shore*. Saint Paul: University of Minnesota Press.

NINE | ADIRONDACK SPRING

Beehler, Bruce M. 1978. *Birdlife of the Adirondack Park*. Glens Falls, NY: Adirondack Mountain Club.

Carson, Russell M. L. 1973. *Peaks and People of the Adirondacks*. Glens Falls, NY: Adirondack Mountain Club.

DeLuca, W. V., et al. 2015. Transoceanic migration in a 12-gram songbird. *Biology Letters* 11: 20141045.

DiNunzio, Michael G. 1984. *Adirondack Wildguide: A Natural History of the Adirondack Park*. Elizabethtown, NY: Adirondack Conservancy.

Hudowalski, Grace L. Editor. 1970. *The Adirondack High Peaks and the Forty-Sixers*. Albany, NY: Peters Print.

Keller, Jane Eblen. 1980. *Adirondack Wilderness: A Story of Man and Nature*. Syracuse, NY: Syracuse University Press.

Ketchledge, E. H. 1967. *Trees of the Adirondack High Peaks*. Gabriels, NY: Adirondack Mountain Club.

Master, Larry. 2015. The Saranac Lake Christmas bird count: a 60-year record of winter bird populations in the central Adirondacks. *Adirondack Journal of Environmental Studies* 20: 67–86.

Peterson, John M. C., and Gary D. Lee. 2008. *Adirondack Birding: 60 Great Places to Find Birds*. Saranac Lake, NY: Lost Pond Press.

Wyckoff, Jerome. 1967. *The Adirondack Landscape: A Hiker's Guide*. Gabriels, NY: Adirondack Mountain Club.

INDEX

North on the Wing

Vireo: Bell's, 135, 136; Blue-headed, 50, 169, 170, 188, 191, 216; Philadelphia, 171; Red-eyed, 191, 194; Warbling, 128; Yellow-throated, 101, 117, 154

Walleye, 158, 180
Walmart, 104
Warbler Big Days, 5–6, 115
warbler life history, 13–17
Warblers, 13; Bay-breasted, 111, 168, 190, 191; Black-and-white, 41, 71–72, 141, 154; Blackburnian, 41, 147, 149, 151, 191; Blackpoll, 8, 41, 136, 137, 147, 218, 220–21, 224; Black-throated Blue, 218, 219, 220–21; Black-throated Green, 41, 50, 111, 149, 150, 194, 195, 207; Blue-winged, x, 4, 113, 114, 117; Canada, 118, 147, 150, 195; Cape May, 146, 168, 190, 200, 202–04; Cerulean, 58, 124, 131, 132, 134, 136; Chestnut-sided, 64, 65, 111, 119, 120, 136, 140, 141, 142, 147; Connecticut, 167–68, 228; Golden-winged, 41, 49, 141, 143, 148, 153–54; Hooded, 56, 57, 60, 71; Kentucky, 56, 65, 66, 86, 119, 136; Kirtland's, 24, 197, 198–202, 204–05; Magnolia, 94, 111, 146, 156, 166, 178–79, 194, 195; Mourning, 142, 147; Myrtle, 1, 137, 163–64, 195, 196, 220; Nashville, 29, 141, 167, 191, 194; Orange-crowned, 146, 171–72; Palm, 147, 167; Pine, 100–01, 117; Prairie, 113, 114, 117, 134–35; Prothonotary, 65, 78, 86, 94; Swainson's, 56, 79, 225; Tennessee, 13–17, 101, 111, 120, 136, 141, 146, 163, 166, 167, 168, 191, 195; Wilson's, 141, 171–72; Worm-eating, 71; Yellow, 111, 141, 146; Yellow-rumped [=Myrtle], 1, 137, 163–64, 195, 196, 220; Yellow-throated, 99, 100. *See also* wood warblers

waterfowl, 176
Waterthrush: Louisiana, 70–71; Northern, 111, 149
Water Ways Festival, 101–02
Waxwing, Cedar, 40, 214
weather and migration, 36–38
West Feliciana Parish (LA), 62
West Texas A&M University, 95
Whip-poor-will, 119, 143, 200
Whiptail, Texas Spotted, 29
Whistling-Duck, Fulvous, 61
White River, 87
Whitney, Tina, 77
Wikelski, Martin, 129–30
Wilcove, David, 115
wilderness, 184–86
wildflowers (various), 178, 190, 194, 209, 218, 220
Wildlife Habitat Council, 103
window strikes, by birds, 127
Winona (MN), 138
Wisconsin River, 131
Wolf, Timber, 142, 149, 153, 177
Woodcock, American, 113, 153–54
Woodpecker: American Three-toed, 160, 174, 193; Black-backed, 151–52, 174, 191, 193; Hairy, 191; Ivory-billed, 75–76, 81–82; Pileated, 66, 82, 213; Red-bellied, 130; Red-cockaded, 88, 100, 105–07; Red-headed, 73, 113
wood warblers, 2, 13; life history of, 13–17. *See also* Warblers
Wren: Carolina, 42; Sedge, 141, 143
Wyalusing State Park (WI), 131–38

Yellowthroat, Common, 20, 25, 134, 136, 141
Yucatán Peninsula, 36, 212

N

Pipestone
Menako Lakes
Badesdawa Lake
Pickle Lake
Nord Road
Sioux Lookout
Dryden
Sandbar Lake PP
Rainbow Falls PP
Pukaskwa NP
International Falls
Thunder Bay
LAKE SUPERIOR
Itasca SP
Sault Sainte Marie
Tamarac NWR
Park Rapids
Duluth
Governor Knowles SF
Crex Meadows SWA
Grantsburg
Luzerne
MINNESOTA
WISCONSIN
Winona
Effigy Mounds NM
Prairie du Chien
Wyalusing SP
Galena
IOWA
LAKE MICHIGAN
MICHIGAN
Missouri
Mississippi
Illinois
INDIANA
ILLINOIS
KANSAS
Pere Marquette SP
St. Louis
MISSOURI
KENTUCKY
CANADA
ONTARIO
Wisconsin